FORCE C

The History of a Lak

CW01667474

30.6.90

Force Crag at the head of the Coledale valley (Findlay)

FORCE CRAG

The History of a Lakeland Mine

By IAN TYLER

RED EARTH PUBLICATIONS
1990

Copyright © Ian Tyler 1990

ISBN 0 9512946 2 8

Published by RED EARTH PUBLICATIONS
7 Silver Street, Marton, Ulverston, Cumbria
LA12 0NQ

Printed in Great Britain by
SMITH SETTLE
Ilkley Road, Otley, West Yorkshire LS21 3JP

CONTENTS

FOREWORD

In the late 1960s I discovered the wonderful freedom and adventure of walking the Lake District fells. During the past twenty years, and frequently accompanied by my family, I have achieved my ambition of climbing and photographing every fell in the Lake District National Park. This pastime, of course, led me to every area of the Lakes and through beautiful, wild and sometimes remote country. I became aware of, and then intrigued by, the numerous mine adits, spoil heaps and derelict mine buildings which I came across on my travels. My interest aroused, my next aim was to discover all I could about the non-ferrous mines of the Lake District and what is left of our dying industrial heritage.

At this time the only books I could find on the subject were by Postlethwaite and Shaw, although later I was to discover Elizabethan Keswick (W.G. Collingwood) and Elizabethan Copper (M.B. Donald). I began to explore, survey, and photograph every mine I could find within the Lake District National Park and also to compile an individual history of each. Extensive research in numerous libraries and museums, but mainly in the archives in Carlisle, Kendal, Barrow-in-Furness and Northallerton, helped me to locate many obscure mining sites which had long since been forgotten. During these years of discovery I acquired an extensive collection of original artifacts, photographs, plans and minerals, plus a comprehensive personal library and archive. The next logical step seemed to be to write a book and I set about considering which subject I should choose from the vast amount of information I had accumulated. Having seen in recent years the closure and destruction of Carrock Mine and the abandonment of Honister Slate Quarry, I turned my thoughts to the last surviving non-ferrous mine in the English Lake District: FORCE CRAG.

It has, I suppose, become a personal obsession to preserve as much of our industrial mining heritage as I can - together with many of my friends in the Cumbria Amenity Trust Mining History Society - but it is a task which is wholly satisfying. To this end, besides writing this book, I have also opened a small mining museum in Caldbeck. Both these projects are dedicated to those men who worked in the dark, with only candles for light, who drove the ground with gunpowder and picks and shovels, who spent years in the wet and cold, and who clawed the ore from the rock with their bare hands...''The Old Men''. To those men this book is a tribute.

Ian Tyler. April 1990

ACKNOWLEDGEMENTS

Numerous organisations and people have helped me in the writing of this historical account of Force Crag Mine and these I list below.

Lord Egremont, for his permission to use the archive at Carlisle Castle regarding lease dates and mineral returns.
Dr. J.V. Bramley, of La Porte Industries, Sheffield.
Mr. M.R. Smith of the Imperial College of Science, Technology and Medicine, London.
The Wolf Safety Lamp Company Ltd., Sheffield.
The British Geological Survey, Nottingham.
Mr. Corker, of McKechnie Ltd., Widnes, for his permission to use documents relating to Force Crag Mine.
Mr. E. Rutter, of Hexham, engineer at Davidsons Ltd.
Mr. J. Hindmarch, of Braithwaite, for his kind permission to use his collection of photographs and for all the help and information he gave quite freely.
Mr. R. Hilland, ex-miner of Force Crag, for advice and information.
Mr. P. Ratcliffe, ex-miner of Force Crag, for advice and information.
Mr. A. Findlay, for his photographs and developing skills.
Mr. E. Brown, for the computer drawings, plans and photographs.
Dr. E. Day, of Portinscale, for his technical assistance and information about Force Crag.
Mr. N. Thompson, of Carlisle, for his invaluable geological knowledge.
Mr. J. Hodgkins, of Altofts, ex-miner of Force Crag, for information and photographs.
Mr. R. Calvin, of Whitehaven, for his photographs.
Mr. P. Fleming, of Barrow, for his photograph.
Mr. M. Davies-Shiel for his photograph.
Penrith Farmers & Kidds, of Penrith, for permission to use their advertisement.
Tiffen King & Nicholson, of Carlisle, for permission to use their advertisement.
The staff of Cumbria Record Office (Carlisle) for their assistance and help in locating DB 58, the old accounts book of Force Crag Mine.
John Foster and colleagues at Carlisle City Library for the assistance given.

Particular mention must be made to the following who gave so much of their time to explore the mine week after week and in all weathers: Warren Allison, Edd Brown, Alen McFadzean, Paul Witheridge, Anne Danson, and Dave Blundell who introduced me to the delights of Force Crag.

I especially must thank the directors of the New Coledale Mining Company, Anne Danson, Peter Blezard, Mike Sutcliffe and Lindsay Greenbank, for all their help and the consideration they have shown over the years, and to Neil Barnes, of Braithwaite, who comes from generations of miners and who is still carrying on the tradition.

Finally I must thank Jean, my wife for her drawings and for all the time, patience and assistance given in compiling the manuscript.

GEOLOGY

The main vein at Force Crag runs north of east through the Skiddaw slates. The vein tends to dip north at 75 degrees and runs the length of the Coledale valley through to Crummock Water, a total distance of nearly four miles. Force Crag Vein is the southernmost of a group of two or three veins, the other named vein being the Long Crag Vein. The Skiddaw slates vary from striped sedimentary stones to a dense rock which is similar to sandstone. The walls of the Force Crag Vein, particularly the footwall, are well-defined and occasionally show traces of slickensiding. Because of the sound nature of the rock, certainly in the lower half of the mine, little timber support is required, thus rendering the mine safe and economical to work.

The vein content varies according to depth. In the High Force area barytes is the main filling with a little quartz and psilomelane, the lower section of the mine, from 800ft to 1,300ft A.O.D., consisting of barytes, galena and sphalerite, all of which are commercial minerals. Numerous other minerals such as dolomite, chalybite, pyrites, cerussite, stolzite, fluorspar and manganese, are to be found. These, however, have little or no commercial value in the quantities discovered and in some cases have caused problems during the milling and separation of the various ores to be mined.

GALENA (Lead Ore)

A mineral known to the ancients, having been used by the Egyptians who gave it the scientific symbol of the scythe. The early Romans were also aware of its ease of use and applied it to water ducting and roofing. Its later uses were for leaded windows, pewter pots, acid batteries and armaments.

BARYTES (Barium Sulphate)

Barium was first prepared by a Mr. Davy in 1808, its name been derived from the Greek ''baros'', meaning heavy. Through the years it has had numerous commercial uses, one of the biggest being for Lithopone, used in photography, but it has also been used extensively in the manufacture of optical glass, plastic, munitions, paper, chemicals, paint, and as a drilling flux.

ZINC BLEND, SPHALERITE OR BLACK JACK (Zinc Ore)

Discovered by the ancients but not exploited till about the 1720s and then only in small quantities. Much of the zinc mined in the UK in the early years was simply thrown away on the dumps or used to backfill disused areas of the mine. The German, Basil Valentine, was aware of its existence back in the fifteenth century and Paracelsus proved its metalliferous property in the sixteenth century. One of its early uses was in the manufacture of brass; it was later used in the manufacture of wire, plate, and for galvanising iron sheeting.

SILVER

This is found in the galena and when separated yields up to 32oz per ton of lead. Most silver was turned into coins, jewellery and plating. Silver gave quite a handsome bonus to many an ordinary lead mine and ensured survival over many years, a classic example being the old Greenside Mine at Patterdale.

1. THE EARLY YEARS

The looming shadow of Force Crag, a 600ft high cliff of gnarled rock, is split by the spectacular Force Falls which, when in spate, drops dramatically in two magnificent cascades to the valley floor and Coledale Beck. Nestling in the corner of the crag, under the scree shed from the southern flank of Grisedale Pike (2,300ft), is Force Crag Mine. The summit ridge of Grisedale Pike marks the northern boundary of the 760 acre mining sett, while to the west, above High Force Crag at some 2,000ft, lies the grassy plateau of Coledale Hause, marking the extremity of the western boundary. Looking down from the summit of Force Crag, one can see Coledale Beck meandering its solitary way down to Braithwaite village, once the home of the many miners who worked at Force Crag and other mines in the area. The beck creates a natural and indisputable southern boundary to the sett.

The mineral rights and land have been in the possession of the Leconfield Estates of Cockermouth Castle since around 1066, when the land was given by a Crown grant to Ranulph Meschines. Since then it has been retained through the line of descent to the present day.

When German miners came to Keswick in the year 1564 with the Hans Langnauer Company, of Augsburg, many were experienced men and had worked in the copper mines at Schwatz in the Tyrol and Neusohl in Hungary. They were brought to this country because their skill, both in the mining and in the smelting of ore, was regarded as being superior to that of the British miner. Mineral mining in this country was in its infancy and the Germans possessed the ability to separate the valuable minerals, such as gold and silver, from the ore during smelting. Queen Elizabeth the 1st was instrumental in forming the Company of Mines Royal, and it was to her that a substantial royalty had to be paid. The company set to work to mine the veins of ore and set up smelters, one of the biggest being at Brigham, near Keswick, and also mills and batteries for the manufacture of copper items such as pots and pans. Any gold and silver produced was sent to the Royal Mint. As these operations were given the royal seal of approval, the Germans could operate virtually anywhere, regardless of who owned the land. Their leader in this country

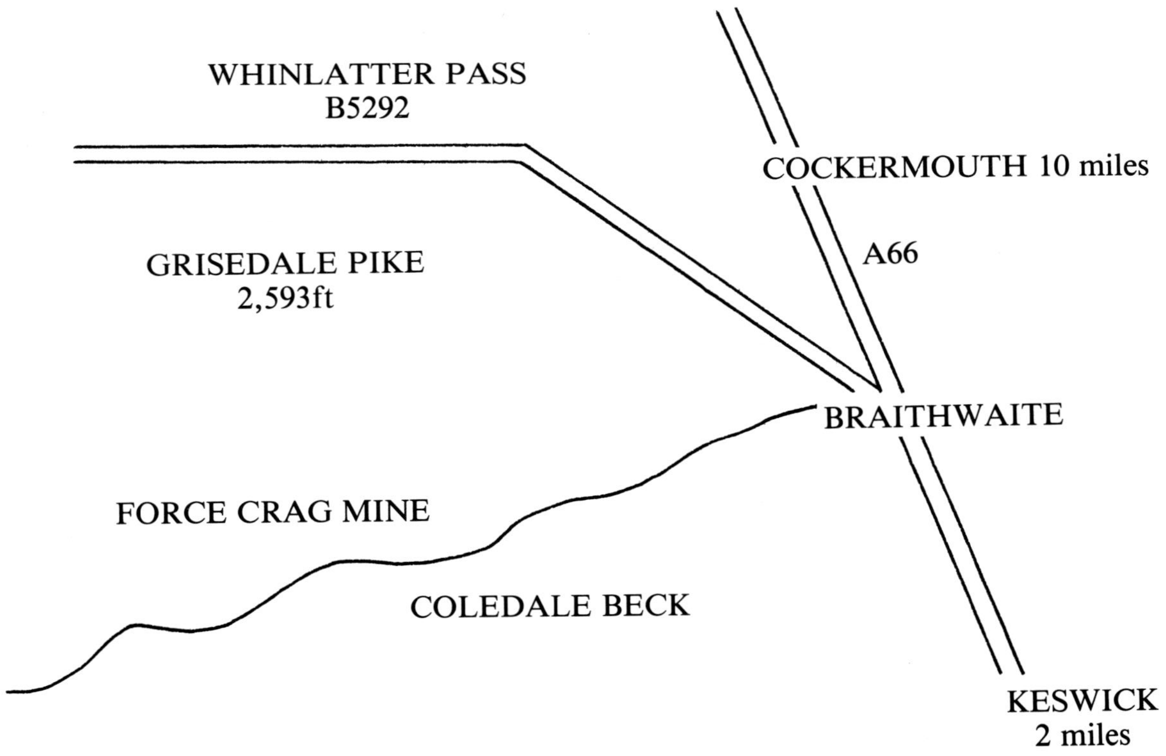
WHINLATTER PASS
B5292
COCKERMOUTH 10 miles
A66
GRISEDALE PIKE
2,593ft
BRAITHWAITE
FORCE CRAG MINE
COLEDALE BECK
KESWICK
2 miles

was one Daniel Hechstetter. He set the men to work in the hills and valleys only a few miles from Force Crag, but it seems unlikely that the German miners ever worked Force Crag itself simply because the minerals in evidence were of little use to them.

It was the gleam of the bright golden copper ore which brought them all the way from Augsburg to mine in the valleys and fells of Newlands, Roughton Ghyll and Coniston - although in Roughton Ghyll they raised some lead ore in 1573. Further to this, however, a report made in 1578 by Thomas Percy, the 7th Earl of Northumberland, stated: "The crust ore at Coledale holds 125 tunne of Lead which holdeth 15oz of Silver per tunne", which would certainly infer that a small openwork had been tried in the vicinity of Force Crag. By this time quite a few miners were living in the Braithwaite and Thornthwaite areas, so it certainly proves that the vein was known even at this early date.

Hidden deep in the ravine carved out by the boiling waters of Coledale Beck is to be found an Elizabethan "coffin level", driven into the north bank of the Kinn, just above Millbecks. The level extends 27 yards to a forehead where a quartz vein is in evidence. The height of this hand-chipped level is approximately 5ft 7ins, the width being 2ft 6ins. Looking from the portal, one can just make out the ramp down to the beckside where once there would have been a bridge for access. All the spoil has been washed downstream. A visit to this level involves a struggle through the wet and boggy ground on the southern side of the beck, this being the only means of approach and then only when the stream is not in spate. The vein in this level appears to be too far west to be the lode worked in the nearby Barrow and Yewthwaite mines, therefore it could well be the end of the Force Crag Vein. The Barrow and Yewthwaite lode cuts through the outskirts of Braithwaite village, near to Millbecks, where this vein has in fact been tried.

So perhaps Force Crag can claim an Elizabethan heritage, for most certainly the existence of its mineral vein was known during that period.

During 1755 prospecting was in progress at Force Crag, initiated by a group of miners who had previously been working Barrow Mine. According to a report at the time, I. Potter, W. Gale and R. Piele made various trials at Coledale Doors but it appears they were unsuccessful as the venture did not progress further.

The established mining sett and its boundary has been outlined previously and from the earliest records little has changed, though occasionally the sett was included with another, by a particular company, to expand or cultivate a larger area. This was the case when on thc 1st of May 1783 John Birkett of Portinscale, John Hodgson, and Edward Barrow, applied for a lease to work Braithwaite, Coledale, Barrow, Barf, Newlands and Windy Hill. It is not clear whether or not any trials were made at Coledale or whether they concentrated on the other parts mentioned in the

Coledale Beck 'Coffin Level' circa 1590

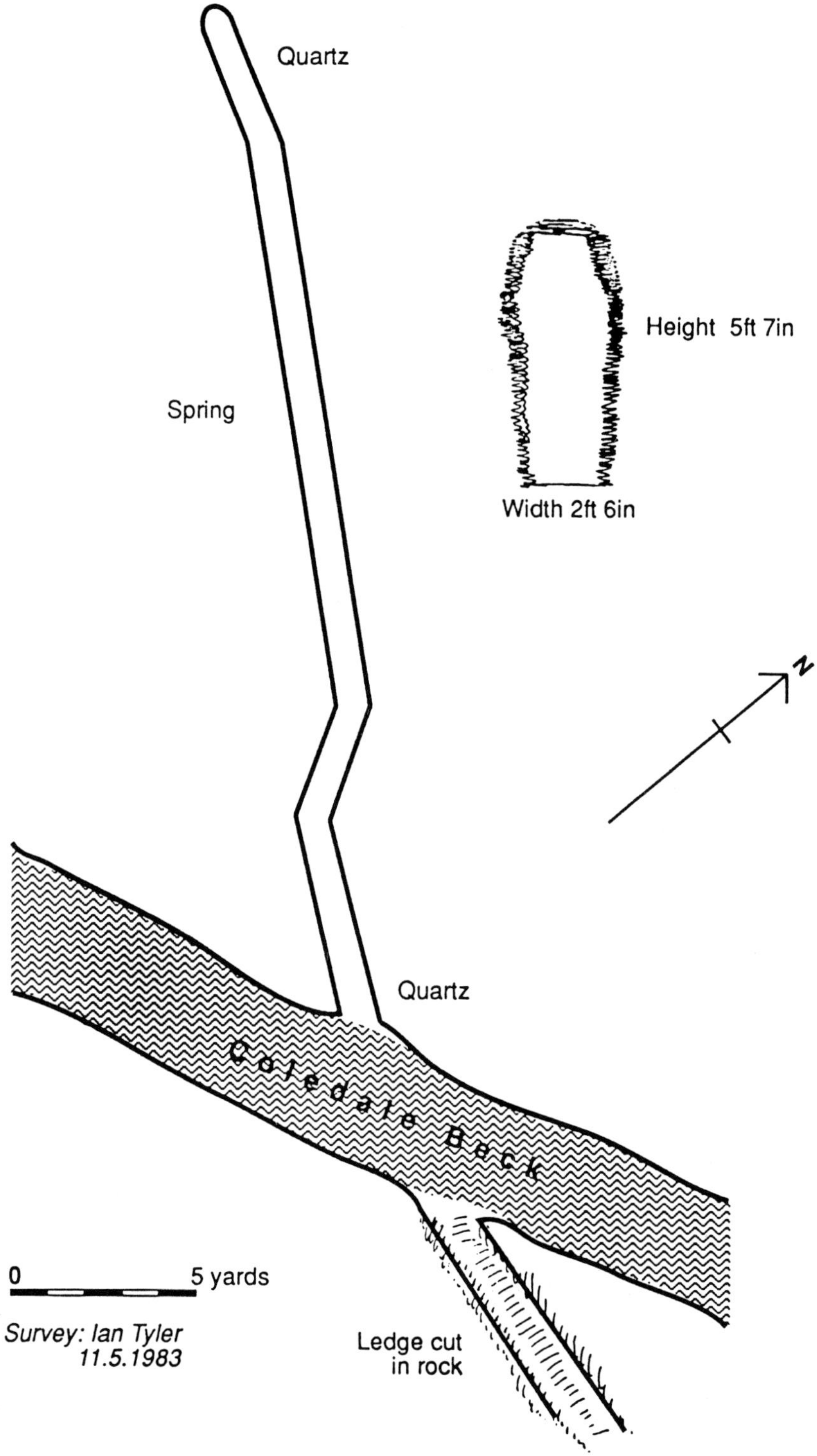

lease, all of which are rich mining areas - but there is no record of royalties having been paid on Coledale.

During this period the village of Braithwaite must have been quite an active place, supporting the mining community who would have worked at Barrow, Barf, Yewthwaite and Newlands mines, all of which were within walking distance. In the village itself were two or three flour mills and a pencil factory which took graphite from Seathwaite.

On the 1st of December 1819 John Tebay, a mine adventurer of Whitehaven, took up a lease which included Coledale Hause, Whiteside and Barf. No money, however, appears to have been spent on exploration although the exposed vein running above the entrance of the modern Force Crag No.3 Level must have been noted for further exploration of a more serious nature. Then on the 4th of September 1826 the same John Tebay took out another lease for twenty-one years but again, for whatever reasons, there was no further development although Mr. Tebay had, in the previous few years, been successfully involved in mining activities at Dalehead, Castlenook, Goldscope and Yewthwaite mines - to mention but a few.

Several years went by before two local men, Mr. H.J. Thompson of Bridekirk and Mr. George Cape of Cockermouth, took out a lease to work for lead on a rental basis, plus a one-eighth royalty. This lease was dated the 27th of June 1838, but again, either through lack of money, or enthusiasm maybe, nothing came of this endeavour and once more the lease was up for the taking. Later in the year more speculators turned their attention to the Force Crag sett and on the 23rd of May 1839 Mr. Tebay was able to sub-let the lease to Messrs John Airey, Emmerson Cowper, and Partners. Initially, the term was for one year with a further nine year option, the royalty payable being one-tenth of the value of all ore sold plus a further shilling in the pound on all lead sold. There was also a covenant obtaining on this lease that two working miners were to be kept in employment for the duration of the lease.

It would seem that the first proper trial was made in the present No.3 Level, where the vein has been exposed by the stream gouging out the gully between the crag and the flank of Grisedale Pike. Work was also started on the erection of a small dressing plant. This would have consisted of an ore washing and sorting floor with a hand-dressing area, possibly covered in as a protection against the weather. The mine itself was being opened up as the dressing floor was under construction but evidently progress in all directions was too slow for Mr. Tebay. On the 1st of September 1839 he put it to the directors that he wished to withdraw from the venture and leave them to continue without him. By this time a waterwheel had been erected at the dressing floor; fortunately, the expense and effort had not been in vain because on the 7th of September 1841, ore weighing 11.74 imperial tons was sold for £15.22 per ton, a sum totalling £183.98.

"HAND-DRILLING"
J. TYLER.

The ore was transported by horse and cart from the mine to Braithwaite and then on to Maryport harbour at a cost of £8.50. From there it was taken by sea to Liverpool at a further cost of £7.99. This method of transporting the ore from its origin at the mine was to continue for many years with little change. The royalty paid to the land owner, Major General Wyndham, was £16.70, leaving the company a balance of £150.79.

The mine was further developed with a crosscut, now known as the No.2 Level, driven to come under the original level. The crosscut was driven a distance of 60 yards before it cut the vein, then, after a short drive east, it continued west along the vein and so under the No.3 Level. As soon as this position was achieved, the miners commenced to stope the vein. By way of a regular progress report General Wyndham was informed of the mine's development. By the 16th of September 1842 the company had already spent a total of £1,500 on building the road up to the mine from Braithwaite, a distance of two miles, and setting up the mill. This was a considerable drain on the company's capital and a request was made that the royalty payments be suspended until the mine was in full production. This request was met with some understanding as the royalty payment was lowered from one-tenth to one-twelfth, thus easing the situation slightly.

It is difficult to be sure of the original dressing floor site but the most obvious place appears to be down beside the Coledale beck, just below the entrance to the modern Zero Level, where the existing road crosses the stream. Here are the remains of a cobbled floor area; there is also the residue of crushed mineral in evidence. Water from the major catchment area was brought to the site by leats, the main stream being dammed just above the mill. The site now lies under the modern slimes ponds, although an old leat is still distinguishable on the south bank. The company also built a large dam under Eel Crag to ensure a steady supply of water during the summer months.

Another mill site is in evidence directly underneath the crag waterfalls, near a sheepfold, where again another cobbled area is to be found along with the remains of building foundations and a set of rusty crushing rollers. However, most of the spoil here is barytes and at this early time the mineral being sought was lead. Although this site was certainly used later, these points would seem to indicate that the original dressing floor was the first mentioned.

The mine began to produce a regular amount of ore and the royalty payments were again adjusted, on this occasion to one-fifteenth. The development work forged on and a rise between No.2 and No.3 levels was completed. This, which later became known as Hall's Rise, effectively aerated the mine and made the movement of ore much more efficient.

The ore was carted via Braithwaite at an average cost of 22p per ton then on to the seaports of Maryport and Whitehaven for transportation to Liverpool. The lead was sent out in wooden casks which were charged to the purchaser at 22p each. Regular customers were the Fallowfield Smelting Company, in the North East, Walker Barker of Liverpool, and Mather Co., also of Liverpool.

Content with the success so far, Mr. Emmerson Cowper requested from General Wyndham that the southern boundary of the sett be extended. This, however, brought an immediate outcry from a Mr. Flintcroft, a director of the Keswick Mining Company, who was at that time exploiting the Cobalt Mine on Scar Crags and who insisted that an intrusion onto the land on the south of Coledale Beck would interfere with his operations. The Keswick Mining Company intended building a smelter beside the beck as soon as the correct formula for the smelting operation had been arrived at, and this would be beneficial to all, including the owners of Force Crag Mine. Based on this argument, the lessees were adamant that none of the land on the south side of the beck could be relinquished, and indeed nothing in this direction ever did transpire.

As can be seen from the table, the first few years of operation had produced a steady tonnage of lead ore and the future prospects of the mine looked good. However, the lease was about to change and new developments were afoot.

The 21st of June 1848 saw the birth of a new company, although some of the old directors and shareholders were retained. The main shareholders in this new company were:

Emmerson Cowper of Alston (miner)
Jacob Walton of Alston (miner)
Watson Dowthwaite of Keswick (shopkeeper)
John Cowper of Alston (mineralogist)

The new lease was for twenty-one years and the royalty one-fifteenth, the specified minerals on the lease being lead, copper, tin, iron, zinc, antimony, sulphur, barytes and manganese. After the ore was taken out of the two levels and dressed it was sent on its journey to Lock and Blackett, or to Shield and Turnbull of Liverpool. The mineralisation in the sole of the lower level (the modern No.2) was looking good so the directors decided that a new level should be driven underneath to prove the vein at depth. In 1849 the No.1 or "Great Level", as it was to be called, was started and driven with great urgency and vigour as the company were keen to exploit the new ground. By 1850 the No.2 Level had been driven a total distance of 120 yards with No.3, the original trial, in 100 yards to the face. The next few years were hard and unrewarding for little ore was produced as all the effort was being poured into driving the Great Level which, by 1853, and after four years relentless

This once busy place is probably the oldest mining site on the Force Crag sett. Worked until quite recently, the entrance to the No.3 Level is now abandoned and partially filled with scree. (Tyler)

effort, was in a distance of 200 yards from the portal. It was an expensive drive, costing up to £9.00 a fathom. The level was cut to a good size for the period, being just over six feet high and about four feet wide, a good "Horse Level" size. The Great Level was going to be the main haulage level for many years to come.

During the year of 1854 a considerable amount of money had been expended on the driving of the new level, so it was decided that more capital expenditure could be justified to build a new mill. The mill was to be sited below the Great Level adit so that the company would be ready to capitalise on the large quantity of lead which would be taken out in years to come. A thirty-foot waterwheel was erected and three leats dug from beneath the southern waterfall to provide enough water to keep the machinery working. The original dam under Eel Crag was also used and the site of this is still obvious today. The new mill comprised a crushing plant, smithy, and other buildings necessary for the efficient milling of lead ore. This mill site is beneath the crag on the south side of the stream, adjacent to an old sheepfold, where a large cobbled area can still be seen. The mill received ore mostly via No.2 Level. From the tunnel entrance it was taken down an inclined track to a small plateau about 120ft directly below No.3 Level. Here, ore was tipped into a crude hopper and hand-broken, then transferred to a smaller hand-dressing floor before being taken down to the mill. The remains of this area can still be seen today, with the exception of the inclined track which has been either swept away or covered by scree.

The Great Level had been driven some 200 yards though had not connected with the No.2 Level as it was not yet underneath. This position would be resolved after a few more fathoms driving when a rise could be put up to effect better ventilation. The miners had been complaining of the poor conditions at the face; the use of tallow candles and the exhausting physical work of continuous hand driving with heavy sledge hammers and thirty-inch drill steels, and the constant banging and noise of the gunpowder charges going off hour after hour, must have stretched the men to their last breath. It must have been heaven to come out into daylight and smell the sweet fresh air of the Coledale valley. Conditions became so bad that the company built a fan which blew air along earthenware pipes to the face, thus making the work tolerable.

In January 1855 the freezing weather and snow affected the working of the mine. In these times nearly all machinery was driven by the waterwheel which, of course, froze up; the mineral was so frozen in the hoppers it could not be moved and consequently no ore could be milled or washed. With the spring came a flurry of endeavour to make up for lost time. The No.1 Level (the Great Level) was pushed forward and was now 300 yards in, and 20 yards from the face the first really good run of ore was discovered. A stope was begun and at last the miners were rewarded for their efforts of the seven years it had taken to come this far. By 1856 the No.1

Level was a hive of activity and, at the stope mentioned previously, thirty tons of good ore had already been taken out. One of the miners from Braithwaite had dedicated so much time to driving the stope and rise that the area - Horn's Rise - was named after him (the remains of Horn's Rise, which did eventually connect with No.2 Level, can still be seen on entering what is now the No.4 Stope in No.1 Level. The rise can be seen in the roof of the stope, some 100ft up).

The stoping of the No.1 Level had coincided with the completion of the mill, which now had a four head stamp battery for crushing the ore. A full washing and processing plant was also in operation, the total cost of all the work being in the region of £1,200.

The mine was expanding and new customers were attracted, one of these being Messrs M.J. Cookson of Newcastle who received ore by rail via Arkleby station near Wigton. In fact it was M.J. Cookson who bought the entire production of 29.50 tons in 1857. During 1858 and 1859 the mining company continued with its exploration and production, the Walker Parker Company of Liverpool purchasing 95 tons of lead for which they paid £1,595.75.

By 1860 the No.1 Level had been driven forward a length of 400 yards. During this period six men were responsible for all mining activity whilst four men, assisted by a few boys, were kept busy in the mill. Although the demand was falling, ore was still being produced and the company had a stockpile of at least 32 tons. The vein they had been pursuing had "nipped in" and the once promising show of ore had been lost. This, coupled with the declining market, was a bad sign and development work was suspended. To keep the mill going the company decided to re-work the old tips, and from this exercise a total of two tons of ore was recovered. However, this was insufficient to reduce the running costs and keep the mine going. The future, though, looked good in another respect.

The Keswick and Penrith Railway Company had been forging its way towards Braithwaite and was expected to be open for mineral and freight traffic by 1865. The railway would speed the ore to market far quicker than the traditional horse and cart; it would also create a bigger customer catchment. But its arrival was not to be soon enough to halt the downward trend in the price and sale of lead. It became impossible to keep the mine going at a profit and so, sadly, Force Crag closed in 1865.

Early 1866 brought prospective lessees to visit Force Crag Mine. These men were John Postlethwaite, James Postlethwaite, John Hall and Joseph Straughton. The ravages of two winters, and vandalism, had taken their toll on the mine. The buildings were wrecked and the level entrances had all collapsed. Undeterred by what they saw, the four men agreed on a take-note of one year, commencing on the 5th of March 1856, with a royalty of one-fifteenth for lead and one-thirtieth for barytes. Work commenced immediately on clearing out the No.1 Level to bring it into workable order, but when the work had been completed nothing further was done and the partnership, for no apparent reason, was quickly dissolved. Later in 1866 a lease was drawn up for Messrs Hull and Parkinson but this was not executed. The mine had, up to now, been operating for nearly thirty years for lead but a new company was about to be formed with the sole intention of mining barytes with lead as a secondary pursuit.

Production figures for 1841 to 1847

	Tons (Imp)	Value	Royalty
1841	11.70	£183.99	1/10
1844	41.00	£513.50	1/12
1845	58.00	£889.35	1/12
1846	25.25	£372.29	1/12
1847	61.75	£831.63	1/15
	197.70	£2,790.76	

Production figures for 1848 to 1863

	Tons (Imp)	Value	Silver Oz
1848	43.00	£519.50	
1849	27.00	£376.60	
1850	7.90	£114.55	
1856	30.10	£421.40*	
1857	29.50	£516.25	176
1858	47.00	£788.75	264
1859	48.00	£804.00	270
1860	80.00	£1,329.00	350
1861	34.00	£490.85	178
1862	12.50	£180.12*	
1863	14.45	£208.22*	78
	373.45	£5,751.24	1,316

* estimated cost.

Ore was supplied to various companies, the largest purchasers of ore being Walker Parker Co. of Liverpool, M.J. Cookson Co. of Newcastle, Shield & Turnbull Co., and Locke & Blackett Co.

2. FROM LEAD TO BARYTES

The new company was formed by three local gentlemen, John Hall of Braithwaite, a mining agent, James Postlethwaite of Brandlehow, also a mining agent, and Joseph Straughton, a druggist from Cockermouth. After a certain amount of legal wrangling the lease was eventually ratified on the 10th of April 1867. Lord Leconfield's lease requested a royalty of one-fifteenth for lead and one-thirtieth for barytes, on top of which was a fixed rent of £20 per annum. The new company drove the No.4 Level, at High Force, on the large, exposed vein of barytes at the top of the gully above No.3 Level. The No.4 Level was named "Postlethwaites Level", presumably after James, one of the directors. The miners set to work with the intention of cutting the vein at depth, as there was a good showing of ore on the surface at O.D. 1,720ft. This would ensure "backs" of 76 feet of easily accessible ore, once the level was far enough in.

The vein, initially, was big and easy to work, barytes being won within the first few days. However, problems were soon encountered in transporting the ore down to the mill, there being no road, or much of a track, up to High Force. The miners themselves made a track up the gully, leading directly from No.3 Level up to No.4 Level, and this provided an easy and quick descent; but the ascent, even for the tough and hardy miners who would no doubt have to carry new drill steels at the start of each shift, was an arduous and grinding slog. For the transportation of heavy gear and timber for supports, the route was impossible. These were transported by pony or packhorse along the diagonal track which can still be found to the south of the waterfall. This track is still used today by walkers on their way to Coledale Hause and beyond.

By the end of April 1867, the first month of opening, 159 tons of barytes had been produced and sold for 75P per ton. To cope with production a grinding wheel was installed to crush the barytes (a grinding wheel was unearthed in a recent dig on the site). The wheel possessed the capacity to crush two tons of mineral a day with three men on hand at the mill to ensure its smooth running. The mill, sited on the north side of the beck and directly under the crag, would have been of a very elementary nature for barytes is easily extracted and processed - because of its friability - and

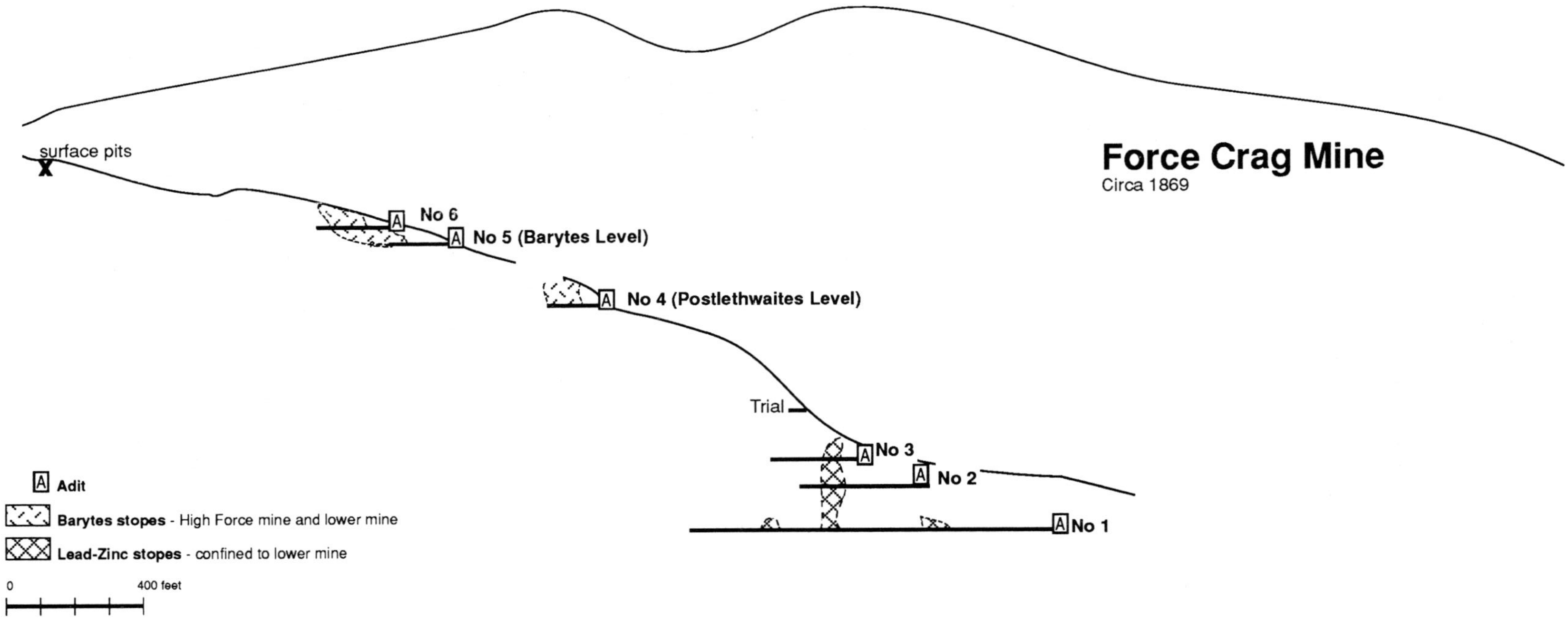
Force Crag Mine
Circa 1869
surface pits
No 6
No 5 (Barytes Level)
No 4 (Postlethwaites Level)
Trial
No 3
No 2
No 1
A Adit
Barytes stopes - High Force mine and lower mine
Lead-Zinc stopes - confined to lower mine
0
400 feet

does not require the sophisticated treatment of separation.

A water leat can still be seen contouring the fell below the cascade of the southernmost waterfall. This leat ensured that a steady flow of water was maintained for the waterwheel which provided power for the crusher. The new company was in a somewhat frustrated position as the ore being won from the No.1 Level was a mixture of lead and barytes and, according to the existing reports, they had no method of treating or milling the lead. From this, one can only surmise that the mill vacated by the previous company, which had been sited only a few yards away, had either been sold, wrecked by storms, or vandalised.

Another leat traverses from the main waterfall of Force Crag, just above the No.3 Level horizon. It seems likely that this was used in conjunction with water flowing down the gully stream to flush the ore down wooden flumes to the hand-dressing floor at O.D. 1,053ft, and so on down to the barytes mill at O.D. 925ft. The stumps of the supports for the wooden chutes can be seen on the fellside. Ore mined from No.2 Level was delivered to the hand-dressing floor along an inclined track running from the level mouth.

By 1869 the mine was progressing well, the No.5 Level being driven along a rich vein of barytes. 96.25 tons of barytes had been sold and this left a stockpile at the mill of 260 tons. In No.1 Level a 2ft wide vein of barytes was being stoped out and was producing something in the region of five or six tons a week. The three men blasting a rise up in the stope complained bitterly about conditions as they had cut an underground spring, which meant they became extremely wet and cold whilst working. It should be remembered that these men wore little or no protective clothing, not even a helmet. Their footwear consisted of wooden clogs, worn without socks, but filled with straw for warmth, so if the straw became wet it could be easily replaced. Sometimes this straw was used to ignite the gunpowder when blasting. Most miners wore a fustian jacket, cord trousers and a woollen shirt, with only a soft hat to protect the head. Lighting, of course, was provided by tallow candles which burnt reasonably well in dry areas but did not survive too well in the damp. Careful positioning of the candles was required when blasting otherwise the miners would be plunged into total blackness, as the flames would be extinguished by the shock of the explosion.

The No.1 Level drive was in some 400 yards and, despite the ventilation fan being in use, the air was not too good at the face or in the stope - although the miners knew that once their rise broke through into No.2 Level things would be much better. The necessity of having to mine the mixture of lead and barytes was an annoyance to the miners because they were obliged to tram the unwanted lead ore to the mill. This was achieved via an inclined track which was put down to the mill as soon as the level started to produce. The track has, since those days, vanished under tons of scree. The unwanted lead was a nuisance to the company as there were no facilities to treat it and capitalise on its value.

Meanwhile, in No.2 Level, forty tons of barytes had been produced from the big rise ascending to No.3 Level, which had become known as Hall's Rise, presumably after the director of that name. The No.3 Level itself was standing idle, awaiting the connection of the rise from below, while in No.4, Postlethwaite's Level, the vein which had once looked so promising had "nipped in" and no further mineral was taken out. The level is much the same today as it was then, with the angle-iron rails still in situ. This is an indication of the faith they had put in the level as rails were a costly item and were laid only in levels which were of great importance. It is surprising that they were never lifted and transferred to another part of the mine.

Above, the No.5 Level, which was later to be known as "The Barytes Level", had penetrated a hundred yards. This was driven directly on the vein which had already produced forty tons of ore. The vein was also proved to the west of Pudding Beck by a series of surface trials at O.D. 2,050ft, these being the most westerly workings within the sett. A small trial was also made on a vein striking east towards Force Crag on the lower reaches of Whiteside mountain at Whin Ben, O.D. 800ft, in Gasgale Gill.

The company were well pleased with their progress, the prospects being excellent. Barytes was in demand, stocks were good at the mill, and orders were on hand. During 1870, 442 tons of barytes were milled and sold.

In order to achieve a pure white product a bleaching plant was installed. The men working the grinding wheel had increased its efficiency and were capable of turning out four tons per day. The directors were quick to realise the potential of the mine and decided to float the company to raise more capital so that they could explore and expand the mine to its full potential. The new company was formed on the 8th of August 1872 and was called The New Force Crag Mining Company Limited, some of the previous company's directors being retained. The senior directors of the new company, with their capital investment, were:

P. Coulthard	£500
John Straughton	£500
James Bolton	£250
William Ashton	£125
Iona Sandwith	£375
Tiffin Bell	£250
	£2,000

Thirty-one other shareholders purchased shares of various denominations from

"THE SHOT FIRER".

£15 to £120, and when all was banked the total invested came to £3,580. The new company was able to benefit immediately from the goodwill generated by the previous company which, at the time of the changeover, had a well established customer file. Listed below are the names of the major customers, many of whom were supplied on a weekly basis.

J.A. Saddler	Middlesbrough	Cleveland Chemicals
Goadsby & Co.	Manchester	Dry Salters
Clay Ltd.	Liverpool	Paint Works
Taylor Ltd.	Liverpool	Paint Grinder
Branthwaite Ltd.	Workington	Paper Mills
Furness Tar Products	Ulverston	
Edmon Holt	Manchester	Chemist
W. Holt	Salford	Broughton Paint Works
E.W. Johnson & Co.	Newcastle	

During this period of working, the rise which was being blasted up from No.2 Level broke out to surface and became known, in later days, as the Old Shaft. Indeed, shortly after this rise was completed the full vertical connection was made from No.1 Level, via No.2 Level, and up to No.3 Level. This undertaking, as mentioned before, was known as Hall's Rise, and rose some 265ft vertically from No.1 Level, providing excellent aeration to the lower workings.

The Barytes Level, No.5, was driven on into a 10ft wide vein of barytes. It appeared that the miners had hit the jackpot! Consequently, the company started to make extensive plans for further development and at a board meeting the following points were decided:

1) For some time now the ore had been transported by sub-contract from the mine to Braithwaite and this amounted to a constant drain on resources. It was decided that the company would construct a horse-drawn tramway down to Braithwaite to alleviate the problem.

2) The company would rent a derelict flour mill and drying kiln in Braithwaite for crushing and drying the barytes. This, in theory, would allow a twelve-month working cycle, as snow, frost, and freezing water would not be as much of a hindrance being so much further down the valley.

3) In order to speed the transportation of ore from the High Force workings (levels No.4, 5, and soon No.6), wooden troughs would be bolted to the crag face

so that the ore could be tipped or "slushed" down to the valley below. The ore was in fact transported down the gully from high above No.3 Level in wooden troughs using water from Long Crag to provide the necessary lubrication to move the ore efficiently. The flumes, on arriving at No.3 adit, were then directed down the fell to the mill. Some of the supporting pins for the flumes can still be seen in the gully; also still visible is the line of wooden stakes which supported the flumes on the final section down to the mill. It is also possible that a similar system was used to transport the ore from the higher workings of the No.5 and No.6 levels. Prior to this, all ore would have been transported by mule or packhorse.

Early in 1873 the No.6 Level was driven to intersect rises already being put up from No.5 Level, where a total of five miners were stoping out the ground. As can be seen from the production tables, over a thousand tons of barytes had been sold in the previous two years - the mine was flourishing! The directors were so proud of their achievements that they invited the directors of E.W. Johnson and Co., of Newcastle, to visit the mine, all expenses paid. The visit took place on the 7th of July 1873 when the visitors were no doubt informed, first hand, of the company's modernisation plans. The cost of the trip, including hotel and rail fare, was £7.26.

The plans laid down previously at the board meeting were put into effect by Mr. J.M. Woods, a civil engineer, who surveyed the land for the inclined tramway to Braithwaite, and whose account for the job was £6.81. An outside contractor excavated the incline for the track and was paid a total of £356 by the mining company. The sleepers were purchased from the Cockermouth & Penrith Railway at a cost of £43.60 and of course, the most important item of all, the iron rails, were purchased in various lots from Messrs Whitewell, of Stockton on Tees, for a cost of £624.31 plus a carriage charge of £25.70.

The tramway was completed with great speed, taking just one year, and was a marvellous achievement. It was two miles in length and required a tremendous amount of excavation as it entered the ravine near the terminus at Braithwaite, where it ended in a trestle construction and hopper. The trestle was known locally as the "Pier End" but, alas, it has now totally vanished. The course of the railway can still be followed, cutting the contours of the Coledale valley at a steady angle, just beneath the mine road. If traced along its length, the remains of old wooden sleepers can still be discovered.

A Horse, purchased to pull the wagons of barytes to the mill, cost the company £25. A field was rented for £5 a year, along with a stable in the village for £1.75.

The flour mill and drying kiln, used for processing the mineral, were situated on the south side of Coledale Beck and to the front of the church. The lease was taken up on behalf of the company on the 4th of August 1874, with a rent of £20 per annum.

Dr. Messrs. E. W. Johnson & Co: Newcastle.

				£	s	d
1873						
Septr.	8	To 10 Tons No 3. Barytes @ 37/6	7	18	15	"
		" Bags – – – –	7	1	5	"
"	16	" 10 Tons " " @ 37/6	7	18	15	"
		" Bags – – – –	7	1	5	"
"	23	" 6 Tons Stone Barytes @ 20/- – –	7	6	"	"
"	25	" 10 " No 3 " @ 37/6 –	7	18	15	"
		" Bags – – –	7	1	5	"
"	30	" 4 Tons " " @ 37/6 –	7	7	10	"
		" Bags – – – –	7	"	10	"
Octr.	4	" 5 Tons " " @ 37/6	7	9	7	6
		" Bags – – – –	7	"	12	6
"	21	" 6 Tons " " @ 37/6	7	11	5	"
		" Bags – – – –	7	"	15	"
"	26	" 4 Tons " " @ 37/6	7	7	10	"
		" Bags – – – –	7	"	10	"
				104	"	"
Novr.	30	To Amt as per day book	8	100		
Decr.	26	" do	8	81	2	6
1874 Janr.	29	" do.	9	90		
Feb.	28	" do	10	94		
Mch	14	" do.	10	50		
				415	2	6
Apr.	22	" do	10	78		
May	26	" do	10	46		
June	19	" do	11	50		
				539	2	6
July	14	" do	11	45	5	
Aug.	28	" do	11	50		
	31	" Balance		81	14	6

Dr. Revenue

Month	Day	Particulars	Fo.	£	s	d
		Forwd.	✓	627	3	10
		To Amt. owing Rail Co. for Carrge	9	66	19	10
July	31	" do to General Balance a/c		386	14	.
			✓	1080	17	8
Septr.	4	" Wages	9	34	2	2
	8	" Gas Company for Coke	.	16	2	.
Octr.	13	" J Muse Salary etc	.	118	9	7
	15	" J Straughton Wages	.	20	17	.
	20	" Jonan. Hodgson Corn		5	8	"
Nov.	3	" Lindsay Joiner		2	"	"
	6	" J Straughton Wages	.	35	8	4
		" Do Rent	.	4	10	.
Dec.	3	" Farrer do	10	5	.	.
		" J Straughton Wages	.	25	13	7
1875 Jan	2	" do do	.	24	19	.
	15	" Rent & Royalty	.	14	15	4
Feby	5	" J Straughton Wages		30	16	3
"	23	" Jonan. Hodgson Corn	.	8	4	"
March	5	" J Straughton Wages	.	21	12	.
April	3	" do do	.	27	7	6
	5	" Wilkinson Blacksmith	.	3	"	"
	26	" Kirby	.	6	3	..
May	7	" J. Straughton Wages	.	35	19	5
	21	" J. Straughton & Others Purchase	.	"	"	"
	26	" do Salary	.	12	10	.
	31	" C K & P Railway Carriage	.	73	18	5
June	3	" Jonan. Hodgson Corn		7	14	.
	4	" J. Straughton Wages	.	26	15	10
July	2	" Lord Leconfield Royalty	,	13	16	8
	2	J. Straughton Wages	.	34	0	11
		Carr over	✓	538	16	6

Dr Force Cragg Mine

Date		Particulars	Fo.	£	s	d
1877						
Sept	1	To Amount expended to date (exclusive of purchase) See Journal, folio	18	973	7	5
Nov	26	To Cash Armstrong for Timber	20	2	10	–
1878 March	31	To Horse Keep A/c Jo. fo.	2	12	6	4
	"	To Wages A/c	2	254	14	3
	"	To Rents & Royalty	2	29	19	10
	"	To Repairs & Renewals	2	27	19	8
	"	To Incidental Expenses	2	7	17	8
			£	1308	15	2
1878 April	1	To Amount expended to date – balance brot down		975	17	5
	"	Wages A/c	5	102	4	4
		Rents & Royalty	5	47	10	.
		Incidental expenses	5		8	6
				1126	.	3
Aug	31	To Amount expended to date – balance brot down		975	17	5
		To Musgrave for Castings } transferred from Miscellaneous		20	–	
		To Blackburn for Timber	6	15	–	–
				1010	17	5
1879		To Balance brought down		1010	17	5
Aug	31	" Wages	8	156 ~~144~~	18 ~~18~~	8 ~~2~~
		" Rents & Royalties	8	54 ~~52~~	1 ~~5~~	7
		" Repairs & Renewals (add Blackburns)	8	17 ~~10~~	10 ~~15~~	11 ~~5~~
		" Incidental Expenses	8	3 ~~1~~	8 ~~2~~	11 ~~4~~
		" Horse Keep	8	1	2	6
		" Coke	8	4	3	4
1879 Aug	31	" Blackburn for Timber	7	10	11	.
			£	1258	14	4

Dr Profit & Loss

Date		Particulars		Fo.	£	s	d
1878 March	31	To Force Cragg Mine	To folio	2	97	13	9
	"	To Miscellaneous Items	"	2	27	14	-
	"	To Bank Interest & Commission	"	2	177	5	4
	"	To Bag account	"	2	71	18	4
	"	To Directors' Expenses	"	2	15	19	10
	"	To Barytes A/c loss on stock	"	2	43	13	6
	"	To Secretary's Salary	"	2	12	10	
	"	To Auditors' Salaries	"	2	3	3	–
	"	To Discounts		2		11	11
				£	450	9	8
1878 April	1	To Balance bro down			364	19	8
Aug	31	" Force Cragg Mine		5	47	19	4
		" Miscellaneous Items		5	54	10	
		" Bank Interest & Commission		5	29	4	1
		" Secretary's Salary		5	12	10	
		" Auditors do		5	1	1	
		" discounts		5	1	5	
					511	9	1
Aug	31	To Balance bro: down			493	19	7
	"	To Bag Account discrepancy in stock			77	2	3
					571	1	10
1879	"	To Balance brot down			516	11	10
Aug	31	" Bag a/c		8	10 (~~36~~)	6 (~~2~~)	3 (~~9~~)
		" Bank Int & Com		8	46	15	10
		" Force Cragg Mine		8	134 (~~128~~)	4 (~~18~~)	11 (~~6~~)
		" Directors Expenses		8	38 (~~3~~)	6 (~~8~~)	9
		" Audit		8	3 (~~2~~)	11 (~~10~~)	0
		" discounts		8	1 (~~1~~)	10 (~~9~~)	7 (~~5~~)
		Printing Stationery &c		8	13	12	4
		Secretary's Salary		8	37	10	-
				£	793	11	0

(It has been rumoured that some of the more unscrupulous local millers would sprinkle a little barytes into their flour to increase its weight). This mill, however, had to be demolished in the early 1900s because of its dangerous condition.

Dividend paid to main shareholders during 1875

	Investment	Dividend
John Straughton	£500	£30.00
Tiffin Bell	£250	£15.00
J. Sandwith	£375	£22.20
P. Coulthard	£500	£30.00
W. Ashton	£125	£7.50
J. Bolton	£250	£15.00

The remaining shareholders were paid various amounts from £1.80 to £14.40, which added up to a total dividend of £289.20. Wages for miners varied of course, but on average their monthly wage account totalled £36. Mr. Straughton's salary was around £29 per month and the company secretary received £12.10. The price of barytes by 1876 was still fairly stable, with the best white bringing £1.45 to £1.85 per ton, stone barytes, which was the middle grade, bringing £1 per ton, and the lower grade, gritty barytes, fetching 36p per ton. All the barytes was despatched from the mill in hessian sacks. Customers were charged a deposit of 3p on each, which was credited to their account on the return of the sacks.

Production over the three years previous to 1875 had been in excess of 1,500 tons. 1877 saw one of the biggest production years with the company selling 588 tons of barytes. The reserve stocks were held at different places. Deposited at the mine were 244 tons, valued at £175.80, while at the low mill at Braithwaite were a further 33 tons, valued £26.85, with a further 17 tons at the kiln store, valued at £20.25. The mine itself was still being developed. No.6 Level had been driven forward and a shaft sunk in its sole, very nearly connecting with the No.5 Level.

It was during this year that an eagle-eyed clerk at the Leconfield Estate office noticed that the details of the lease were not being executed properly, the aberration being that the company had, since 1872, failed to pay the statutory annual rent of £25. One can well imagine the uproar in both offices. This was an ill wind for it meant that a further call was made on the shareholders to raise the capital. Few accepted, however, and many resigned themselves to forfeiting their investment, although one or two new people did enrol. But this was not enough to stop the downward spiral.

The sale of barytes continued but the price began to fall, as can be seen during 1878 when 436 tons was sold for £305.20, an average of 70P per ton. The final straw was the horrendous winter of the same year which prevented any mining being done,

"LIGHTING THE FUSE"

although the mill at Braithwaite was operational. 1879 saw the price of barytes falling even further; consequently the mine was becoming unprofitable, causing the shareholders and the directors to become increasingly worried. John Muse, the mine manager, with his many years experience, was powerless, having thousands of tons of good ore in the stopes for a market that had collapsed within a year.

John Straughton, as the head of the company, had the painful task of writing to the mineral office at Cockermouth Castle, on January the 19th 1881, to inform Mr. Glutton, the mineral agent, that the mine had become totally unprofitable and had ceased production. He explained that over the past fourteen years of working the mine, the company had spent in excess of £7,000 on equipment and development, of which he had a personal investment of at least £800. However, the Cumberland Union Bank Ltd had foreclosed on four of the directors: Messrs Ashton, Bolton, Blackburn (a new director since 1876) and himself, from whom they had demanded £1,000 immediately. On these grounds, and pleading extenuating circumstances, he requested they be relieved of the binding contract of the lease. It is probable that this request was granted with the proviso that the site be cleared.

It was on New Year's day 1882, then, with sleet and snow in the air, that William Postlethwaite - the son of James Postlethwaite - and two mine agents from Braithwaite, Thomas Jones and Dixon Barnes, set off on the long haul up the road to the mine. All experienced mining men, their mission was to see if Straughton's company had left anything worthwhile at the site, for use, should they decide to take out a lease.

What a scene of disaster and dereliction greeted them! In the short time the place had been vacant the mill roof had been blown off, some of the main structural walls were down, and the timbered entrances to levels No.1 and No.2 had completely collapsed under oceans of scree. When they trudged up to No.3 Level entrance it was only to find their way barred by a huge snowdrift. Although it would probably have been obvious that it would be pointless to continue on up to No.4 Level, nevertheless, they did. Maybe it was William, wishing to inspect his father's level, who urged them on but, alas, after the long struggle up the gully, the level could barely be seen because of the snowy conditions and the scree which had enveloped the entrance. Disillusioned and despondent they began to retrace their steps down to Braithwaite, no doubt ruminating on the wisdom of selecting such a day for prospecting and eagerly returning to the warmth of their firesides, a tot of rum, and to discuss the future prospects of the mine.

January the 2nd 1882 brought yet another visitor to Force Crag Mine, this being none other than Lord Leconfield's bailiff, no doubt checking to see if anything could be salvaged. Whilst there he was met by the previous day's visitors. Nothing fruitful came of the meeting and in the cold light of day the venture was not pursued. Sadly, the mine was allowed to fall into complete disrepair and by 1885, after an appalling winter, when the mine was next inspected, it was found that all the low levels were completely run-in. No.4 Level entrance could not be located at all and levels No.5 and No.6 were likewise completely scree-filled. The mill itself was totally wrecked and worthless. What dreadful desolation had descended upon a place which had been the site of such busy and fruitful industry, where men had been employed in hard but honest labour which, indeed, was no more than they had asked. The mine was about to enter one of its darkest periods and little or no interest was shown with the exception of an enquiry during 1886 by a Mr. McDougal, who applied for a lease. This lease, although not executed, was to be the last show of interest before the turn of the century.

Ore Production figures 1867-1879

	Tons (Imperial)	Value
1867	600	£420.00
1868	213	£130.70
1869	96	£67.20
1870	442	£287.30
1871	572	£400.00
1872	504	£352.80
1872 (zinc)	600	£600.00
1873	230	£161.00
1874	381	£266.70
1875	700	£320.30
1876	442	£355.90
1877	568	£568.10
1878	436	£305.20
1879	188	£131.60
Total barytes	5,372	£3,747.20
Total zinc	600	£600.00
Total production	5,972	£4,347.20

The zinc production figure for 1872 is somewhat suspect as the mine was on line for the production of barytes. The barytes figure given for 1867 includes two amounts, some production figures stating that 441 tons were produced in 1866 and 159 tons in 1867. This, in effect, would mean that production commenced before the lease was signed. It is possible that some form of dispensation was given in view of the fact that the gentlemen had a take-note on the property which had barely expired. The production figures are, I believe, accurate although the cash returns for 1871, 1874 and 1879 are estimated values based on ore prices of the day.

3. INTO THE 20th CENTURY AND ZINC

New legislation was about to be introduced which would greatly improve working conditions for the ordinary miner. Until now he had no redress against such things as the lack of basic safety precautions in the mine, no pay while off work through sickness or injury sustained on the job, and even, sometimes, eviction, for many men lived in houses owned by the company. Any prolonged period of illness, therefore, meant no wages and, consequently, the rent could not be met. Conditions were generally poor, entailing long working hours, often in dismally uncomfortable situations - and low wages. It is, however, fair to say that not all mining companies laid down such harsh conditions; by all accounts the coal miners faired much worse than the metalliferous miners. New legislation known as Postlethwaite's Code was introduced for the safety of the workers. It was made law, in 1896, that at least two hours before the commencement of a shift, the mine manager or his deputy would inspect all shafts, manways and working areas to ensure they were safe for the miners. Another safety factor introduced was that stagings on manways should be no more than eighteen feet apart, this distance having now being increased to thirty feet.

The transition to the 20th century went totally unnoticed at Force Crag, in fact it was to be several years before anything at all happened there. The first signs of interest came from a G.G. Blackwell & Son who applied for a lease to work the mine in 1905 but the application lapsed and came to nothing. Then a much more genuine enquiry came from two gentlemen by the names of Mr. Joseph Frost Lobb, a mining engineer from Portinscale near Keswick, and his partner Mr. Thomas Crawford Dennison, an engineer from Middlesex. It was on the 18th of December 1906 that a take-note was agreed for a period of one year. The terms were that a fixed rent of £25 per year was payable with a royalty of one-twentieth on all ore raised. This must have been somewhat optimistic in view of the disastrous state of dereliction which the mine was in and the length of time it would take to get things back into working order.

The first priority was to open up the levels and inspect the workings for it had been nearly twenty-five years since anyone had been underground and levels No.1 and No.2 were standing deep in water which was held back by tons of scree. It is, in fact, still possible today to see the water mark on the level walls. It took all year to clear out the levels and prepare for the recommencement of mining operations.

Confident in the future success of the mine a full lease was drawn up and signed for a thirty year term by Messrs Dennison and Lobb, on the 29th of December 1907. The details and covenants of the lease were as follows: a £100 rent would be paid for each of the first two years then £200 for the third and subsequent years. The royalty would be paid at one-twentieth of the selling price of ore sold and shipped. A minimum of six miners were to be employed; open stopes, air shafts or dangerous openings were to be securely fenced and any damage to crops or agricultural land, roads or buildings, would have to be rectified. Pollution of streams was to be prevented by the construction of settling ponds. A surety of £20 was to be deposited but any negligence or failure to comply with the terms laid down could result in the lease being revoked. The new company was called The Cumberland Mines Limited.

During the period of the take-note a lot of ground work was done. No.1 Level was retimbered and the entrance secured, while the level itself was driven forward; in No.2 Level the men were driving on a lead vein while No.3 was being investigated for ore. Machinery was brought in and a small mill erected outside No.1 Level. This soon proved to be inadequate and a mill was designed with the capacity to process the zinc ore properly. This was erected immediately below No.1 Level on the site of the present-day mill.

During this period the workforce was kept to a minimum and in 1906 there were only three miners. By 1907, however, numbers had risen to six working underground and seven on the surface. Just as the new venture was getting underway Mr. Dennison resigned from the board, though Mr. Lobb was joined by two new directors, Mr. Fred Meakin and Mr. H. McKenzie, a new agreement being drawn up on the 5th of November 1908.

Production of zinc ore resulted in ten tons being sold while the workforce increased to a total of fifty men, twenty-four of these working underground and twenty-six on the surface. It is likely that the large number involved on surface work was by reason of the erection of the new mill, which, for its time, was of modern construction.

The mill was built of stone for the main part though wood was used for some walls and the roof. Power was supplied by two gas engines which also provided compressed-air for the newly arrived air rock drills. It should be noted that for years Force Crag Mine very rarely employed a smith on site, as was the custom in most mines, but retained the local blacksmith, Joshua Dobson, in Braithwaite, whose smithy was at "Peter House" in the centre of the village, now a private residence.

LOBB'S MILL FLOW SHEET, CIRCA 1907

ORE FROM No.1 LEVEL

ORE TIPPED INTO HOPPER

BARREN ROCK AND OVERSIZE SORTED BY HAND

20 x 10 INCH STONE BREAKER

CRUSHING ROLLERS

TROMMEL SCREENS

GREEN'S JIGS

DIESTER SLIMES TABLES

BALL MILL (Possibly a PEBBLE MILL)

BUSS TABLE

ORE FOR SMELTER

This mill was operational just prior to the installation of the Elmore Flotation Plant.

Work continued on No.2 Level where a good showing of lead and zinc was to be seen; this ore was drawn and fed to the mill, which had a capacity to handle a run of mine ore of 75 tons per day. Unfortunately, problems were experienced in separating the barytes, zinc and lead. The difficulty lies in the fact that these minerals have a similar specific gravity and even today, with all the technical knowhow available, these three minerals, when found together, present a challenge to the people working the mill. In spite of all these trials and tribulations the company succeeded in producing, by 1909, 9 tons of lead, 35 tons of barytes, and 140 tons of zinc. Shortly after this success, for reasons unknown, the company went bankrupt. The causes could have been many: over-expenditure on the mill, the fact that the lead and zinc vein they were following had pinched out, or perhaps the separation process was inadequate. But bankrupt they were and The Cumberland Mines Limited ceased to trade as a company in 1911.

This was a bitter blow to the local men who had assumed their livelihoods were secure. But the spirit of adventure beckoned and more than one Braithwaite miner left family and friends and made the long journey to Australia and the Broken Hill lead and zinc mine. One of these emigrants was William Hindmarch, whose family still live in Braithwaite village. Tragically, another local man, Thomas E. Turner, was to be less fortunate. On transferring his services to Threlkeld Mine he purchased a motor bike to travel back and forth. Shortly after moving he was involved in a fatal accident, on Braithwaite bridge, when his bike collided with a van.

New legislation governing metalliferous mines was brought in during this year. The new Act recommended the following: that no boy under the age of twelve years and no girl or woman of any age be employed to work underground; that no boy between thirteen and sixteen years of age be employed on a Sunday, on weekdays before 5am or after 9pm, on Saturday again before 5am or after 2pm and for not longer than ten hours a shift; that the next shift to be at least twelve hours after the one prior and that the full working week was not to exceed fifty-four hours.

These new laws would have been introduced in time for the latest company at Force Crag, which was formed by the directors of the previous concern, Mr. Joseph Lobb and Mr. Thomas Dennison. The name of the new operation was The Coledale Syndicate, the lease being signed in 1912 for a period of thirty years, at a royalty of one-twentieth with rent set at £100 for each of the first two years and £300 for the third and subsequent years. The company signed on eighteen men straight away and set seven to clearing and timbering No.1 Level and another two on clearing No.2. The rest were put to work repairing the top dam under Eel Crag, constructing a header tank and sluice gates and then to laying the nine inch iron pipes down the 600ft of fell to provide water for a new Pelton wheel which was to supply power for the mill and compressed-air for the rock drills.

During this period the mill underwent extensive alterations, the most interesting and important of these being the installation of an Elmore Flotation Plant, a brand new system of separating the ore after crushing and so maximising the yield of mineral brought out of the mine. In fact, this plant was one of the first to be installed at a British mine.

The first recorded patent for an ore dressing process similar to flotation was taken out in 1860 by William Haynes. Haynes found that on mixing certain fats or oils with powdered ore, the sulphide mineral particles showed a marked affinity and could be removed by washing the mixture with water. In 1886 and American woman, Mrs. Carrie Everson, patented a process for mixing pulverised ore with various fats or oils and certain acids and salts, and for water separation of the oiled sulphides to be floated away from the unwanted gangue particles. However, no successful commercial process was discovered until 1898 when F.E. Elmore patented his system and applied it to the concentration of gold-bearing chalcopyrite ore. In the bulk oil flotation the separation of the oils carrying the sulphides from the remaining material made use of the buoyancy of the large amounts of oil used. The possibility of using gas bubbles to increase the lifting power of the oil was not recognised at first. This was discovered by Messrs C.V. Potter and G.O. Delprat in 1902. At about the same time a Mr. M.A. Froment recognised the advantage of using gas bubbles in an oiled pulp, utilising the selective action of the oil for the sulphides and the selective action of the gas for oiled particles. His patent was bought up by Minerals Separation Limited, an English company which played an important role in the subsequent development in oil flotation. The Elmore vacuum process, patented by F.E. Elmore in 1904, was installed and operated successfully in a number of plants. It was carried out as a continuous process in an enclosed apparatus, gas bubbles being produced in an oiled pulp by creating a vacuum over the pulp. The reduction of pressure served to release dissolved air and other gases; the bubbles attached themselves to the oiled sulphides and carried them to the surface. It was a modification of this system which was installed at Force Crag Mine, the apparatus being housed in a large wooden tower.

The workforce was increased and a total of forty men employed, twenty-eight underground and the remaining twelve on duty in and around the mill. During the year of 1913 a total of 32 tons of lead was produced and from this 640 ounces of silver recovered. The flotation plant and mill were modified slightly, for the dark clouds of war were beginning to roll over Europe and it was obvious that the home market for all minerals would soon be at a premium. In No.1 Level a crosscut had been driven east and discovered a large deposit of ore; the working in this deposit became known as the No.1 Stope. Mineral reserves at this time were estimated at something like 40,000 tons in No.1 Level alone, containing an estimated 6% zinc, while at the same time zinc was also being won in No.2 Level.

In 1914 the company were laying plans to expand the mine in the hope of capitalising on the need for the minerals they had to offer. It was decided by the board to drive a new major level a hundred feet below the No.1 and some way to the east. This level, a crosscut driven from beside the mine road, was to become known as Zero or No.0 Level. A good deal of money was poured into the venture for, on striking the vein, the new level would certainly give the company reserves for a long way west, and at a good depth, under No.2 and 3 levels. Much timbering had to be done as the ground was exceedingly poor. The manager, Mr. G.F. Wallace, was, to say the least, extremely frustrated by the progress of the new level for after months of work no vein was to be seen. The level soon acquired the nickname of the "Unsuccessful Level". The large amount of money plunged into this venture had been allocated at the expense of the rest of the mine, which had been virtually neglected. The anticipated demand for ore failed to materialise quickly enough, so with all the capital spent on the mill and the new level, and the returns not forthcoming, the company was unable to continue.

In time, though, as the war continued to rage in France, and Germany's blockade and submarine threat began to strangle the country, the Ministry of Munitions turned to the home front for raw materials. Force Crag's potential was realised by the ex-manager, Mr. G.F. Wallace, who naturally knew the mine well. Wallace formed a partnership with a Mr. C.E. Currie, and between them, and with the help of the Ministry, they formed Braithwaite Mines Limited with a capital of £15,250.

This company pushed levels No.2 and 3 on into new ground, 180ft and 60ft respectively, and by 1916 further plant was installed to streamline production. The new plant included a Harding Pebble Mill. This was a rotating drum containing large, carefully selected sea-washed pebbles of a very high density. Into this the ore was placed, then, as the drum rotated, the ore was crushed. Three separation tables were also installed. These were fluted tables on which the pulverised ore was deposited. A constant stream of water washed away the dross, leaving only the heavier pure mineral. Also included in the new plant was a larger crusher which increased the capacity of the mill by three tons per day. The miners were putting in long hours to help the war effort, and for their dangerous job they received just £1.32 per week.

Production through 1917 was good with stopes being worked in both No.1 and No.2 levels. A total of ten men were working underground and six men on the plant. However, things do not run smoothly in the business of mining for very long. It was during this period that the nine inch pressure pipes, leading from the top dam to the Pelton wheel, ruptured. One can well imagine the velocity of water coming from 600ft above and the problems involved in repairing the pipe. As a result, production

The construction of the Cumberland Mines Limited mill, circa 1906, under the direction of Mr. Joseph Frost Lobb. Force Crag's present day mill has been grafted onto Lobb's original structure. (Blezard collection)

The mill, circa 1916, shortly after the mine was taken over by Wallace and Currie's Braithwaite Mines Limited. The wooden tower at the right of the picture housed Joseph Frost Lobb's Elmore Flotation Plant. (Hindmarch Collection)

Peutnay Porter, Ned Graves, and Dr. Horten Smith at the original entrance to the Great Level (No.1 Level), circa 1916. (Hindmarch Collection)

Ned Graves and Peutnay Porter, having a smoke in the sun. Shortly after this picture was taken, Porter was killed in a Derbyshire lead mine while attempting to rescue his trapped son. (Hindmarch Collection)

The Force Crag mill, circa 1916. (Hindmarch Collection)

Miners in Milkhouse Rise, No.3 Level, circa 1916. (Hindmarch Collection)

Detail of the interior of the mill, circa 1916. (Hindmarch Collection)

Ned Foster with horse and cart, circa 1916. Note the stockpile of white barytes in the open shed. (Hindmarch Collection)

was lost for a while.

As the pipes were being relaid and secured, Zero Level was being driven and was now in some 300ft, with the vein still proving elusive. It was around this time that a sump was sunk in the sole of No.2 Level with a view to making connection with No.1 Level, but after sinking twenty feet the project was halted. The mine manager, a Mr. Peutnay Porter, who had, incidentally, recently returned from his mining adventures in the Yukon, received a telegram informing him of a serious accident at the big Millclose lead and zinc mine in Derbyshire. The telegram informed him that his son, who was a miner there, was trapped underground after an explosion had caused the workings to collapse. Peutnay set out immediately to assist in the rescue but, tragically, he was killed in the rescue bid and his son was saved.

The mill was slightly updated with the introduction of two suction gas engines and a new Silex Ball Mill, installed to grind the concentrates from the jigs. This method differed from the old Pebble Mill in so much that the pebbles were replaced by 4-5 inch diameter steel balls, which proved to be far more efficient. In fact, this type of machine is still in use today. The jigs separated the light country rock from the heavy mineral by means of gravity and pulsating water. From the Ball Mill the concentrates were fed into the Elmore Plant and so onto two new concentrate tables. Mill and mine were now working efficiently. The compressed-air rock drills were making the drilling of the shot-holes a much quicker process, compared with the laborious hand-drilling method, and therefore speeding the stoping and driving. With the mill being on line it was well able to cope with the increased output and was capable of handling over thirty tons a day.

In 1918, with the Great War nearly over, there was a sharp fall in the price of zinc. The total workforce at the mine at this time was twenty-five men which, as the management was well aware, was a major draw on finances. Some men were laid off but enough were kept on to work the stope in No.1 Level and try Zero Level again. A further thirty feet was driven but the level still proved to be useless and so it was once more abandoned.

The end of the war came and the company was notified by the Ministry of Supply that the subsidy on zinc was to be removed. This was a bitter blow for the company had concentrated its efforts towards zinc extraction, more or less ignoring the barytes and lead. Although zinc, in 1920, was still bringing £11 a ton, the price was plummeting. At their height during the war mineral prices had been:

	Zinc	Lead
1914	£23.30	£19.90
1915	£66.65	£22.86
1916	£68.45	£30.95
1917	£52.17	£30.00
1918	£52.20	£30.10
1919	£42.25	£28.20

Cumberland production figures for the war years

	Barytes	Lead	Silver	Zinc
1914	nil	2,406	14,835oz	6,053
1915	1,888	897	5,791oz	3,788
1916	3,436	839	6,666oz	2,907
1917	3,954	720	6,445oz	2,411
1918	4,998	706	5,975oz	2,510
1919	4,854	1,405	10,959oz	3,787

It was in 1920 that a fatal accident occurred at the mine. It is surprising that in such a dangerous occupation there were not many more serious accidents concerning explosives, ladders, whirling and grinding machinery of all types, not to mention rock falls and moving or rotting timber. Perhaps it is a credit to the skill of the joiners who constructed hundreds of feet of false floors, shuttering, and hoppers, which had to withstand the battering and crashing of hundreds of tons of ore, much of it falling dozens of feet. Some of the credit must go to the skill of the blasters who set the explosive charges so precisely when breaking new ground, and brought down just enough muck so that the tunnels continued at a uniform height and width. Setting charges was always a dangerous occupation, all charges being lit with a slow-burning fuse. Once the fuses had been ignited the men retired swiftly to count the number of detonations from a safe distance.

Whether miner Joss Dixon, of Braithwaite, miscounted the reports, or counted an echo by mistake, one will never know. As he returned to the face to inspect his work a rogue charge exploded and Joss was killed instantly. Accidents of this type were few in the non-ferrous mines of the Lake District, probably due more to the ability of the miners than, as some would say, "good luck".

A year later, in 1921, the mine was virtually at a standstill, although the fruitless drive on Zero Level was again being pursued and was now in 350ft. It was certainly living up to its nickname of the "Unsuccessful Level". The remaining nine miners were employed either in Zero Level or in sinking a new sump in the sole of No.1 Level, 350ft from the entrance. This exploratory work was to try to prove the existence of barytes at depth, as an alternative mineral to zinc. The directors of the syndicate were desperately trying to raise the £25,000 needed for the further development of the mine, and in particular the workings at High Force.

By 1922 the sump in the floor of No.1 Level was down 26ft and a large vein of good quality barytes, 3ft wide, had been cut. A total of 50 tons was raised and Mr.

Wright, the current manager, was well pleased with the strike. However, this showing of the vein and the reported reserves still waiting to be exploited in the High Force were not enough to tempt would-be speculators. The wretched Zero Level was forced another ten feet, making the total drive 360ft, and all that there was to show was a tunnel with no apparent sign of a working vein. The directors, try as they might, failed to win financial support and so the mine closed.

A year later there was a routine inspection of the mine and, once again, and how truly devastatingly, the elements had taken their toll (anyone who has visited this place during the winter months needs no reminder of the overpowering force of the weather conditions which obtain in and around the high fells and valleys. I personally have seen cars physically bounced and moved by huge gusts of wind; I have been peppered with scree blown off banks hundreds of feet away and, on one occasion in 1987, at the entrance to No.3 Level, my fourteen stone, 6ft 5ins frame was lifted completely clear of the ground and unceremoniously dumped twenty feet lower down the spoil heap. I have seen the stream increase from a lucid, murmuring beck to a wildly raging, impassable torrent. In inclement times this is indeed a hostile place). The inspection revealed that, only one year after closing, all the levels had run in with the exceptions of No.3 and No.6, and the mill itself, built by the men of Braithwaite Mines Limited and once a fine example of its type, was in a severe state of dereliction.

It was noted by the mineral agent at Cockermouth Castle that the rent was in arrears to the tune of £392.50. Consequently, the mine was put up for sale in the hope that some cash could be recouped. Meanwhile, the faithful Mr. Wright, the old mine manager, was trying with a friend, Mr. Martin, to raise enough working capital to reopen the mine himself. Both totally convinced of the future of the mine they paid £150 off the rent in 1925 and a further £40 the following year. 1926, though, was the year of the Slump and the General Strike, and raising money for a mining venture was a non-starter, despite their unremitting efforts. Two years passed, by which time the rent arrears had risen to £387. In 1928 Lord Leconfield exercised his prerogative and took over the derelict mine property, and any worthwhile equipment was taken in lieu of rent arrears.

4. CHANGING FORTUNES

Shortly after the mine was reclaimed by the Leconfield Estate it was put back on the market and, almost immediately, the mineral agent at Cockermouth Castle received an application for a take-note. This was from two men who were new to the area, Mr. W.T. Donovan and Mr. Maurice Newbould. Therefore, on the 25th of March 1928, a take-note was agreed for the duration of one year at a rent of £25, with an over-riding deposit on the existing mill and machinery of £50 and an option for a thirty year lease. A company was formed and called The Derwent Fells Mining Company Limited, the directors deciding that the market they would pursue would be in barytes, for this market was fairly buoyant, the United Kingdom having produced some 57,095 tons in that very year of 1928. It was the derelict area of High Force Crag they would attempt to exploit - a serious undertaking, as this part of the mine was the most inaccessible and no work had been done since 1880, some fifty years previously.

The new men were aware of the vein and its course, through exploration of the No.4, 5 and 6 levels. Their plan was first to drive a level some 123ft below No.5, where on striking the vein they would have good "backs" both east and west. The High Force Crosscut, as it was named, was started in good solid rock, just to the east of Pudding Beck, and was cut to a good size, being 7ft high and 4ft 6ins wide, and, initially, being hand-drilled and blasted forward. The company had, no doubt, read of the many problems encountered by previous companies concerning the inaccessibility of the High Force workings, so a small track was forged up the fell by which could be brought the petrol compressors for the air rock drills. The compressors were installed in huts close to the level mouth, the stone, used in some of the retaining walls, being blasted from a small rock face just to the west.

The arrival of the air drills speeded up the driving and, after progressing about 480ft, a string of barytes was cut, this being tried for a short distance both east and west. However, disappointment at the small size of the vein dampened their enthusiasm somewhat, as it bore no resemblance to the massive vein seen in No.5 Level. After careful measuring it was decided to push the crosscut on; their courage

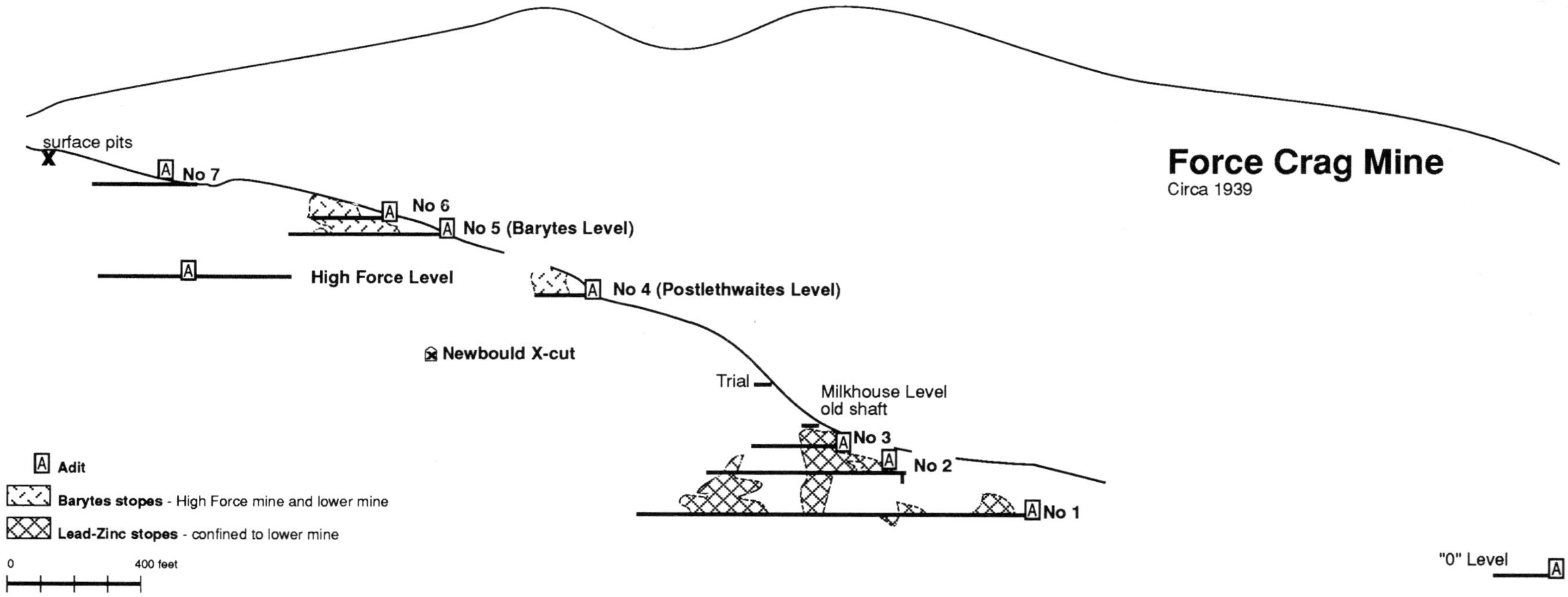

Force Crag Mine
Circa 1939
surface pits
No 7
No 6
No 5 (Barytes Level)
High Force Level
No 4 (Postlethwaites Level)
Newbould X-cut
Trial
Milkhouse Level
old shaft
No 3
No 2
No 1
"0" Level
Adit
Barytes stopes - High Force mine and lower mine
Lead-Zinc stopes - confined to lower mine
0
400 feet

was rewarded for before them, at 525ft from the entrance, and barring their way, was a wall of barytes completely covering the whole of the drive and, as far as they could make out, 7ft to 8ft in thickness. It had taken ten men nearly two years to cut the vein - what a bonanza it was going to be!

In 1930 some 1,750 tons of barytes was won at Force Crag and a royalty paid of £87.50, plus the rent. How the ore was transported down to the mill is cause for debate, but if it was brought down by lorry then the men driving this tortuous road could have performed the stunts for the film ''Hell Drivers'' - only this was for real! Work continued through 1931 with the level being driven east and west to prove the vein. The miners were now seeing their dream come true, with barytes above, behind and underneath. A further exploratory drive was commenced below and to the east of the High Force Crosscut. This was called the Newbould Crosscut, though it was never pursued more than a few yards. In anticipation, the workforce was increased to twelve men and during 1932 and 1933 the mine produced nearly 3,000 tons of barytes, on which was paid a royalty of £150 and a rent of £37.50. Then, for some unaccountable reason, production ceased, although there was still ore there for the taking. One reason may have been that the government of the day had just put a 10% duty on barytes and 20% on lithopone which at the time was the new base product for paint (and the largest consumer of barytes), replacing the old lead base.

The mine closed down once again, although this time not through lack of mineral. By 1936 the production of barytes in the United Kingdom had risen to 73,070 tons, proving that demand was still there. Further figures reflect the growth rate showing that by 1938 home production had reached a staggering 108,000 tons with a further 40,000 tons imported. Of this total amount, 30,000 tons was used for lithopone paint manufacture and 35,000 tons used in the chemical industry, and yet, unbelievably, no-one took on the Force Crag lease when it appears the market was so volatile.

During 1939 a Mr. A.E. Wilson visited Force Crag to make an official report on its condition and its prospects for future development. As usual the findings were not good. The buildings were in disrepair and the only levels accessible were No.3, in the lower half of the mine, and the new High Force Crosscut. The other levels being run-in at the entrances, access to the mine was gained via No.3 Level and the situation discovered was as follows: No.1 Level was in 900ft and was showing four stopes and a flooded sump 26ft deep; No.2 Level was in 540ft and had two stopes; No.3 Level was stoped to the surface and could be reached by the ladderways of Horn's Rise from No.1 Level some 232ft below; Milkhouse Level was a fairly recent trial which had been driven about 20ft on a barytes vein just above No.3 Level, and No.6 Level driven a total of 354ft west and 126ft east. The High Force Crosscut was entered and the vein examined, but levels No.4 and 5 could not be found. Bearing in mind the state of the world at this time one would have thought it prudent to consider a little speculation in the mineral market. However, Mr. Wilson's recommendation was that there was no future and that any attempted venture would probably fail. Little did he realise what was going to unfold here over the next few years otherwise his report might have been a little more optimistic.

Tampimex Oil Products Limited, of Queen House, Kingsway, London WC2, disregarded Mr. Wilson's report and took up the vacant lease in 1939, just at the outbreak of the Second World War. The rent was £200 per annum with a royalty payment of 5p a ton. The mineral they intended to win was again barytes which, incidentally, was now used in the manufacture of explosives and there was certainly plenty of that to be used in the next few years. It appears from the outset that the company had the backing of seemingly unlimited cash resources (possibly from the Ministry of Munitions).

The first job was to establish a good road communication with High Force, achieved by means of a little blasting near the waterfall to widen the existing track. The mill built by the Cumberland Mines Limited back in 1909 was incorporated into the foundations of a new mill, Davidsons of Hexham employing a team of men to install the machinery early in 1940. The new mill and plant was designed by architects and engineers from Whitehaven and Workington. A Mr. Rutter was one of the engineers responsible for installing the plant. All the newly-manufactured parts such as bucket elevators, trommel screens, jaw crusher, crushing rollers, vortex classifiers, concentrate tables, drive shaft bearings and pulleys, were installed fairly easily. The jigs, however, were brought from Threlkeld Mine, near Keswick, and had not been used since that mine closed in 1928. Because of the difficulty presented in their installation, the jigs had to be completely renovated. Launders and chutes were all made on site to fit as required.

Power to work the mill was provided by a 75 horse-power Blackstone engine. Ore was washed, as before, with water pouring from No.1 Level. The level was now a few degrees warmer than normal for the miners had cut a warm spring, a factor which eased conditions slightly during periods of adverse weather.

All this activity had, so far, been concentrated on the low mine but now the High Force was about to receive similar treatment. The road now completed, heavy gear could be transported up the fell. Part of the rock face at the entrance of the High Force Crosscut was blasted out to make way for a compressor house; this was sited to the west of the adit and was constructed from reinforced concrete. To the east of the adit was built the receiving area where ore was sorted by hand and sent down a large concrete chute to the despatching hoppers. From the hoppers the ore was transported by lorry down to the mill.

Adjacent to the sorting area was a small hut which the miners used as a place to eat their bait. And what a wonderful view they would have had on a clear day, being able to see as far as Cross Fell and the Pennines in one direction, and in the other, the Eastern Fells rolling endlessly away to the horizon. Nearby, the massive ramparts of Eel Crag loom over the site, offering a never ending change of mood as the mists swirl across the face. From here the eye travels to the rounded top of Sale Fell and onward to the dark face of Scar Crag and the site of the old Cobalt Mine.

The mill buildings were constructed from local stone and standard red brick, most of the concrete mix being made from tailings from the mill. Sometime in 1940 ore production was commenced in the High Force workings but it soon became apparent that transportation of the ore by lorry was a fruitless exercise. Because of the restricted load capability, the time taken for the journey was prohibitive. Down at the mill there was either a feast or a famine. A regular flow of material was required and the solution appeared to be to erect an aerial ropeway from the mine to the mill. The British Aerial Ropeway Company Limited, of London, was commissioned to implement the scheme. A terminus was erected beneath the hoppers at High Force and from there a two-and-a-half inch steel hawser, suspended from pylons, descended the 740ft - and a distance of some 1600 yards - to the mill. From the hawser were hung the thirty or so half-ton side-tipping buckets which carried the ore. Eventually, it was planned, the ropeway was to extend all the way down to Braithwaite railway station but this part of the project was never completed.

By 1944 a vast amount of work had been done in the upper workings, the High Force Crosscut having been pushed 350ft to the west, along the vein, and 650ft to the east. A shaft was also being sunk in the sole of the level, at the intersection of the crosscut and the vein, and this had reached a depth of 25ft. Over the total length of the 1,000ft level there were no fewer than six main rises, all of which passed through the No.5 Level and up to the No.6, from where there was one main rise through to day. Seventy percent of this ground, which from the High Force Crosscut to the surface covered a vertical distance of 270ft, had already been stoped out. A new tunnel, No.7 Level, was now established and had been driven 270ft but so far no stopes had been cut. The mine produced its largest tonnage of barytes in 1944 when 11,439 tons were raised, the total production reaching something in the region of 28,000 tons by 1945. This vast quantity of ore was removed, literally, by hand, shovelled from the stopes into wooden ore tubs which were then pushed, fully loaded, all the way to the loading bay where the ore was transferred to the ropeway and despatched to the mill for treatment. The wooden tubs were made by Hudsons of Leeds and were the end-tipping type, with a capacity for about . ton of ore. There are still four of these tubs remaining in the level to this day.

Various reports claimed that the ropeway was ineffective and was constantly breaking down or icing up in bad weather, but the company probably accepted this as a hazard of working in so high and remote an area. When the ropeway was inoperable the ore was transported to the mill in four-wheel-drive trucks. Towards the end of the war, in 1945-46, production eased off and only 7,000 tons went through the mill, although development work still continued in the shaft below the High Force Crosscut. Nearly all the drilling for the charges in this part of the mine was done with ''auger'' type drill steels, varying in length from two feet to ten feet, which chewed through the soft rock. The charges were all fired by slow burning fuses, which had to be hand lit, and which proved to be a problem as this was such a wet mine. However, the miners were now using carbide lamps, which gave a much more comforting light and were not as susceptible to being blown out, as were the candles used in previous years.

By 1946 the shaft in the High Force Crosscut was down to a depth of 80ft, the ore being winched upwards in large round kibbles capable of holding around half a ton of material. 50ft down the shaft a sub level, known as the 50ft Level, was driven off. 40ft to the west it penetrated a good pocket of vein stuff while eastwards it ran through good ore for 230ft. The host rock was not as solid as in other parts of the mine and fairly substantial pillars of ore had to be left to support the stope walls. At the bottom of the shaft a level struck out along the vein, east and west, a distance of 150ft. The conditions in the High Force were nearly always appalling as, being near the surface, the water streamed into the mine through the open stopes and ventilation rises, and during prolonged periods of rain the lower level had to be cleared of water. The Newbould Crosscut (at 1,550ft O.D.), commenced by the previous company to prove the vein at depth and laying 80ft below the present workings, would have been ideal as a drawing and drainage adit. However, development of this level was never pursued at all.

Tragically, the grim winter of 1947 began and with all its might and ferocity descended upon Force Crag, bringing with it the worst storms of snow and sleet in living memory. The mine was totally cut off, it being six full weeks before the men regained access - and then only to have their worst suspicions confirmed. On entering the High Force Crosscut they found not only that the shaft was brimming with water but that torrents of melting snow and ice were still crashing into the level from above. They realised that the mine was lost, for without pumping gear, which was not readily available, the shaft could never be dewatered. Consequently, Tampimex Limited relinquished the lease having during their operation raised some 35,000 tons of barytes, spent around £20,000 on the mill and development work, not to mention a further £10,000 on the aerial ropeway, which was dismantled some time later and sold for scrap.

Almost two years were to pass before the La Porte Chemical Company became interested in the mine. La Porte was a large chemical business with offices in Luton,

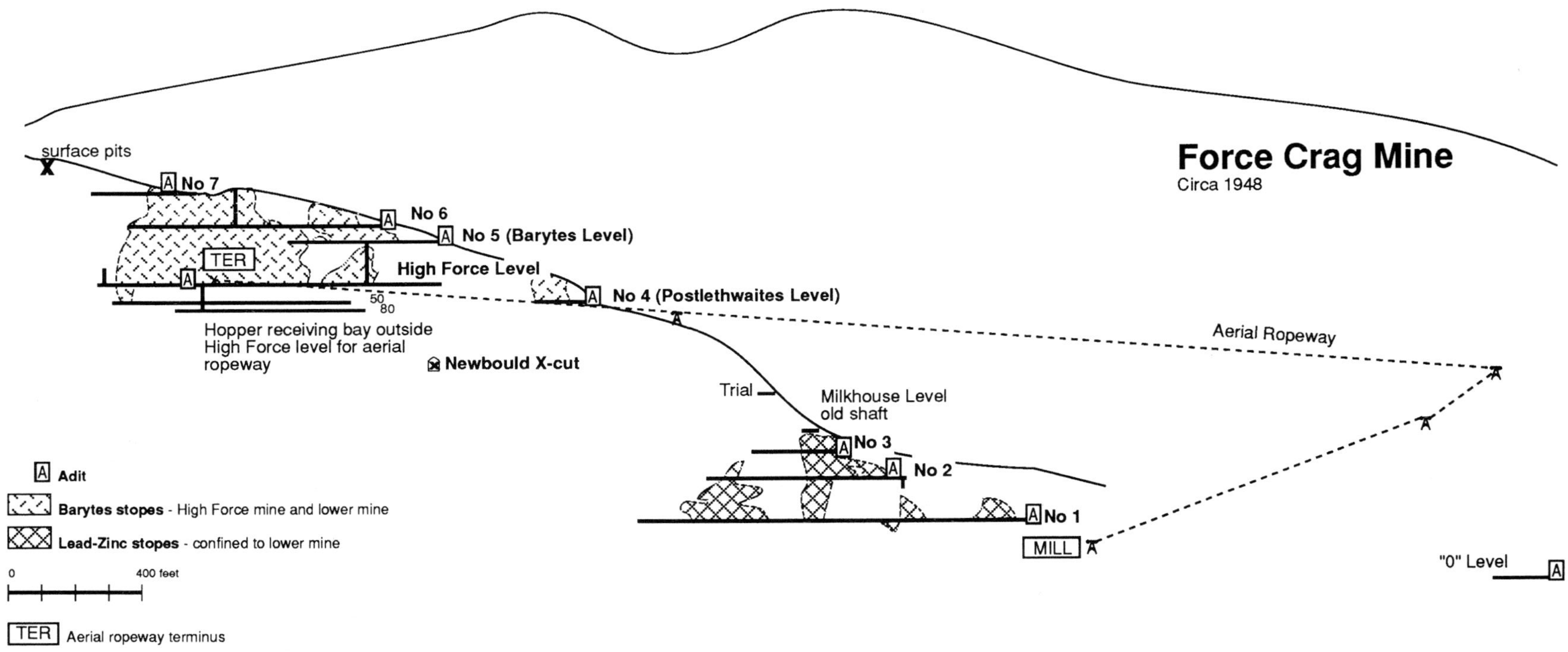
Force Crag Mine
Circa 1948
surface pits
No 7
No 6
No 5 (Barytes Level)
High Force Level
TER
50
80
Hopper receiving bay outside High Force level for aerial ropeway
Newbould X-cut
No 4 (Postlethwaites Level)
Aerial Ropeway
Trial
Milkhouse Level old shaft
No 3
No 2
No 1
MILL
"0" Level
Adit
Barytes stopes - High Force mine and lower mine
Lead-Zinc stopes - confined to lower mine
0
400 feet
TER Aerial ropeway terminus

Force Crag, black above a snowy Coledale. (Calvin)

Ropeway pylon and the snow covered summit of Outerside Fell. (Fleming)

Open stopes high on the flank of Grisedale Pike. (Tyler)

Huts and the loading bay of the aerial ropeway outside the High Force Crosscut. (Tyler)

The loading bay at the head of the aerial ropeway, High Force. (Hodgkins)

Hudson's wooden ore tubs in the High Force workings. (Tyler)

Looking out of the Milkhouse Level. (Brown)

Outside the High Force Crosscut. The miners' bait cabin. (Tyler)

Sheffield and Warrington, complete with its own mining division with resources to explore for minerals. Once the lease was established the surveyors and engineers set to work to decide how best to tackle the mining operation and how to win the barytes hidden within the depths of Grisedale Pike. They were, of course, aware of the mass of ore still in the stopes of the High Force, both east and west of the Crosscut, and that the vein had been proved to a depth of eighty feet below the adit. The same vein was also to be seen in the No.3 Level, some 537ft below the High Force Crosscut, and running to the east. The geologists pondered on whether the vein would remain strong along its length, at this depth, or perhaps nip out. If it did continue in strength it would be a prize indeed! But which way to solve the problems? To drive the Newbould Crosscut forwards and when underneath the High Force workings put a rise up for drainage, would be the most straightforward option, but this would mean re-siting the loading bays and the ropeway terminus which, at this stage, had not been dismantled. The shaft could have been pumped out, which would have been relatively simple, but this would have required a pump to have been left working constantly, due to water continuously pouring in from above. Also, the deeper they went, the more difficult and expensive the pumping operation would become. The reserves of barytes in this area were estimated at 100,000 tons or more. It was just too big an opportunity to miss...so one of the most ambitious mining operations in the Lake District was about to commence.

The plan finally agreed upon was to drive the No.3 Level forward a few yards and then strike an inclined drive of nearly 1,100ft in length and rising in height - over this distance - approximately 325ft. At this point a level would be driven under the flooded 80ft, 50ft and High Force levels, and a rise put up to drain the workings. During the driving of the "La Porte Incline", which was the name given to the project, side drives would be put out to test for the vein as work went forward. The solution found, the benefits all round would be as great as the project itself. The whole of the vein would be proved, the majority of the workings would be self-draining, and ore would be passed swiftly down the incline and trammed out via No.3 Level and down to the mill. The weather would be of little consequence as all the work was being carried out underground where the temperature remains constant at around 52 degrees Fahrenheit, so there would be no problems with freezing hoppers and icing aerial ropeways. The miners, even, could literally walk to work undercover.

It was decided that all the waste rock, which would be created in the driving of the incline, could be disposed of in the Old Men's workings known as Hall's Rise, which came up into the sole of No.3 Level. This was the ideal solution as this stope was now finished and had no workable ore left; in fact, back-filling in this manner would help support the walls and ensure the stability of the mine. It would also mean that the rubbish would not have to be trammed so far. Everything seemed too good to be true!

Driving commenced along No.3 Level; the work went well and soon the point was reached where it was decided the incline would strike off. Here, at what would be the base of the incline, an area was chambered out to give enough headroom for a hopper to be installed, so that trucks in the No.3 Level could be brought underneath and the waste rock tipped in straight off the incline. After the initial 150ft of driving in shaly ground, which required some timber support, the incline entered good solid rock for its entire length. The tunnel was cut to a good size, being 7ft 6ins high and 6ft wide, although these dimensions did vary. The rattle of the compressed-air drills and the crump of the gelignite charges soon became a daily routine for the men at the face. They were able to push forward between 3ft and 4ft with each round of holes fired. But the rock was much harder than at first anticipated and although they were using tungsten-carbide tipped drills, progress was slow.

At a point 280ft up the incline the first side-drive was driven south for a few yards but no vein was cut. The drive, though, was a useful place for storage, a place for the miners to retreat into when blasting and to sit and eat their bait. The incline was pushed on tirelessly and side-drives were tried both south and north, but still no vein of consequence was seen. The company was becoming edgy, for the vein should have been cut by now. The incline was in a distance of 600ft when it was decided to increase the angle of the drive from 30 degrees to about 40 degrees to facilitate the removal of the waste material, which was being scraped down the incline. The miners were becoming frustrated for there was still no sign of the vein, the air was becoming poor, and the continual noise of the charges reverberating down the confines of the tunnel was getting to them. Eventually the company decided that the incline had gone far enough and at this point a level was driven south. The drive was known as the 1100ft Level, this being the length of the incline at this point. A short level was also driven west in another attempt to find the vein, but still nothing was found. The 1100ft level was again driven on, first south, then west, and at last, in an area of disturbed and soft ground, the lode was located some 200ft from the head of the incline.

A rise was put up some forty feet in good solid ground, a short distance from the top of the incline. Then, on the verge of success, the company made a complete turn around in its policy and pulled out of prospective mine development. This was apparently brought about by the falling price of barytes and other minerals, many of which could be purchased abroad. The mine closed without further ado in 1952, the equipment sold off without any mineral having been produced apart from the odd sample which was taken for analysis. The cost of the project is rumoured to have been somewhere in the region of £20,000.

Ore production figures for 1908 - 1946 (Imperial Tons)

	Lead	Silver	Zinc	Barytes
1908			10.00	
1909	9.00	270oz	140.00	35.00
1913	32.00	640oz		
1914	12.00		23.75	
1915*			170.00	
1916*			155.00	
1917*			150.00	
1922				50.00
1930*				1,750.00
1932*				1,500.00
1933*				1,500.00
1941				283.00
1942				3,611.00
1943				9,213.00
1944				11,439.00
1945				7,300.00
1946				3,154.00
TOTAL	53.00	910oz	648.75	39,835.00

* Production figures are estimates based on royalty payments.

5. THE McKECHNIE BROTHERS

Eight long years went by before any more work was done at Force Crag, by which time the mine had fallen into its usual state of dereliction. This was of no consequence to the new owners, McKechnie Brothers Limited, who were a well established barium chemical company with their own mining division, based in Widnes, Cheshire. They had been mining in the Lake District for some years, having been exploiting the barytes veins on the Caldbeck Fells. They were, in fact, still leasing and working the mines of Potts Ghyll and Sandbeds, which they had controlled since 1940. The mine manager there, at this time, was Mr. W.T. Shaw, a well-known local mining engineer who had been involved with many Lake District mines. The Force Crag operations must have been of interest to McKechnies for many years, being in the same business as La Porte Limited. McKechnies were aware of the vast amount of preparatory work done by the previous companies and realised that, with a reasonable cash injection, the mine could become a profitable operation with minimum effort. All McKechnies had to do was locate the vein between the High Force Crosscut and No.3 Level horizon, then bring the mill up to working order ready to receive the ore.

After appraising the situation carefully McKechnie Brothers signed a twenty-five year lease on the 13th of October 1960, at a rent of £50 per year plus a royalty of 22p a ton on barytes and one-thirtieth on lead and zinc. The initial capital injection was £20,000. With this investment the mill was to be put in working order and the High Force levels, the La Porte Incline, and No.3 Level were to be cleared and prepared for the removal of barytes from the High Force. Initially, the 1100ft Level at the top of the La Porte Incline was driven forwards into the vein left by the previous company, though here the rock was rotten and required timbering. Also, the barytes was tainted with manganese and was not as pure as that seen in the lower mine.

Undaunted, the miners commenced to drive a mini incline from the top of the 40ft rise put up by the La Porte company. This would enable the miners to tip the ore from the 80ft Level down to the 1100ft Level. The incline was driven at an angle of

45 degrees towards the flooded 80ft Level. The pilot drill, in the capable hands of Robbie Hilland, soon broke through, and, on its release, the water, which had been trapped since the 1947 flood, poured down the new incline. The workings were left to drain for a few days, and when the deluge had subsided the incline was completed. On making the connection with the 80ft Level, this section of the mine was immediately aerated. Meanwhile, in anticipation, the miners railed the 1100ft Level so that ore - once the connection was made - could be trammed to the head of the La Porte Incline and tipped. The ore would then be scraped down the full length of the incline with the aid of heavy scrapers, dragged by winches, water which ran from the old open workings being used to help lubricate and assist the scrapers (the company which provided the equipment was Pick Rose Ltd). This method of ore transportation is called slushing.

However, some of the problems encountered by the previous occupants soon came to light, one of these being that the transfer of heavy material up the incline to the upper workings was totally impracticable. Also, the ore from the High Force - it was realised - would be man-handled far too many times: out from the hoppers, along the 1100ft Level to the La Porte Incline, down the incline, out through the hopper at the bottom, then into the trucks on No.3 Level. It would then be hand-trammed to the entrance to be tipped into lorries and driven down to the mill. This was totally ineffective and methods of streamlining the operation had to be looked into.

The situation of reserves in the High Force workings was being looked at to ensure a steady flow of ore to the mill. It appears that there were 7,000 tons of ore above and in the High Force Level, with a further accessible 15,000 tons between the 80ft and the 1100ft levels, though some of this ore formed retaining pillars. Towards the end of 1960 the full complement of staff at the mine was fifteen miners, sixteen millmen, four labourers, two office staff and a foreman.

All the miners at this time were working in and above No.3 Level on the development of the High Force. Further to the original plan it was agreed by the Board that the High Force Shaft should be equipped with a power winch for the hauling and lowering of heavy equipment. The road to the High Force was widened with a few explosive charges neatly laid by one Robbie Hilland, a Keswick man, thus ensuring an easier and safer drive up what had become known to the miners as "The Burma Road". The transport used on the road consisted of four-wheel-drive ex-army trucks, ideal for the terrain and able to carry the timber and a variety of necessary mining gear which, in turn, could be lowered down to the 80ft Level via the High Force Shaft. Special flat trucks were used in the 80ft Level on which were set large round kibbles which were hauled up the shaft and out to day, the ore being transported down to the mill by road. This method was time consuming and was soon to become obsolete on completion of the ore-pass, which was being driven at the end of the 1100ft Level to connect through to the 80ft Level. To transfer gear to the head of the La Porte Incline the company devised an alternative scheme which consisted of a tubular steel sledge, about 6ft long, 18 inches wide and about 12 inches deep, and made by Messrs Drysdale and Co. of Yoker, Glasgow, which was dragged up the incline on the slusher cables.

Eventually the mill was in a position to produce, its capacity being in the region of 80 tons a week. On production, the ore was to be sold direct to McKechnie Brothers at Widnes, at an initial price of £5.65 a ton plus a carriage charge at the going rate. The odd shipment did go direct from Force Crag to ICI; however, these shipments were infrequent.

By 1962 things seemed to be looking better with ore coming down the incline from the new ore-pass and the mill beginning to produce - it appeared that most of the obstacles had been overcome. It came, then, as a bitter blow when a visit by H.M. Inspector of Mines resulted in a report which insisted that the La Porte Incline be classed as an unsafe area when the slushers were working the ore, and that the place could not be used as a walkway. This would effectively mean a suspension of work while the recommendations from the Ministry could be considered. The recommendations were as follows: A) that a parallel tunnel be driven, or B) that a conveyor belt be installed from the 1100ft Level to the bottom of the incline, with footsteps and a handrail fitted for the protection of the men. Some sort of compromise must have been reached for a conveyor was certainly never installed; however, the top section of the incline, from the junction of the 900ft Level, is much wider than the lower section, and there is evidence of steps having been cut on the right-hand side, which would have enabled the miners to have walked the incline in safety.

In order to satisfy the Mines Inspectorate, and also to further the development of the mine, it was decided that a series of levels would be driven, commencing with the 900ft and 650ft levels. Once these tunnels had been driven forwards to come under the workings above, twin rises would be put up to the 1100ft Level. One rise would be an ore-pass, the other a climbing way for the men. It was also decided that a major ore-pass and a service rise, of no less than 200ft, would be sited at the junction of the new 650ft Level and the La Porte Incline. Here, ore which had been slushed down the incline from the 900ft and 1100ft levels would be accepted and dropped down the 200ft shaft to an extended No.3 Level. The ore won from the 650ft Level would also be despatched down the ore-pass to the No.3 Level.

The most important and final part of the operation was to drive the No.3 Level forward as quickly as possible so that the 200ft ore-pass and service rise could be put up. This meant that the level had to be extended a distance of 650ft. No time

MILL FLOW SHEET. McKECHNIES 1962

20 TON CAPACITY HOPPER

|

HAND-PICKING BELT

|

13 x 8 INCH CRUSHER

|

100 x 30 INCH VIBRATING SCREENS

|

TOP DECK 5/8 INCH HOLES

|

BOTTOM DECK 1/10 INCH HOLES

|

5/8 INCH MATERIAL TO 18 x 10 CRUSHER

|

ROLLERS

|

SCREENS

|

FOUR COMPARTMENT HARTZ JOY JIGS

|

MIDDLINGS UP ELEVATOR AND OVER SCREENS, MINUS 1/10 INCH MATERIAL WHICH WENT TO DUPLEY PAN AMERICAN JIGS

|

WILFREY TABLES

|

SMELTER

could be spared as this was the key to the efficient transference of the ore. The extra expense of the operation raised a few eyebrows in the boardroom but it was realised that the carrying out of this development would enable the company to search for the deeper section of the vein they had come to work.

The No.3 Level became the priority and the best miners were put on the job. Two of these men were Tom Nicholson and the previously mentioned Robbie Hilland, both local men. The mine manager, John Simpson, was to ensure the work progressed smoothly along with Willie Shaw, the engineer. Initially, the drive was in good solid ground, which slowed up progress somewhat as the rock took longer to drill. When the level had gone in a distance of 150ft it was decided to blast an air vent to link with the incline 45ft above. This rise was put up by Tom Nicholson who, being a big man, cut big rises as all his eighteen stone had to fit into the shaft. The company always got its money's worth from Big Tom. The rise is slightly offset from the side of the La Porte Incline, where it makes its exit, and here the miners have cut a narrow rock channel to prevent water cascading down from the incline onto the men working below. Completion of this rise, early in 1963, saw the lower half of the La Porte Incline, below the 650ft Level, virtually redundant.

During this period little or no ore was being produced. Attention was drawn to the lower levels to check the value of the reserves of ore. No.3 Level was still being pushed on and was in a total of 650ft and had reached the point where the 200ft ore-pass and service rise was to be put up to the La Porte Incline. Up till now, all mucking out had been done by hand but a new battery loco was brought in which could haul six tubs of rubbish. The driver of the loco was Owen Beattie, who was to remain loco driver at the mine for many years.

The 200ft rise was driven by Big Tom Nicholson and Stan Mawson and took three months to complete. From time to time Robbie Hilland also assisted although his main job was heading the drive on the No.3 Level. After 700ft the ground in the No.3 became rotten and had to be arduously and slowly timbered, foot by foot. This was time-consuming work, all the rough timber - which was supplied from the local forest at Thornthwaite - having to be cut on site, then wedged and fitted. Once the timbered section was completed the level resumed in good rock, and at 850ft a nine inch thick vein of pure white barytes was struck. However, fate was to be unkind for this was not the vein for which they had been searching and it soon nipped out. At this point a further 438ft had been driven, making a total drive of 1,288ft.

In March 1963 crosscuts were blasted both north and south from the No.3 Level in a last desperate attempt to find the lode, but yet again results were not encouraging. A barytes string was cut but this tapped a sulphurous spring which filled the level with obnoxious gasses, not to mention gallons of water. This was the final straw, for the miners - by nature a happy bunch of men - were in low spirits, having driven the level in so far but with no result. Conditions at the face, already poor, became intolerable because of the water and bad air. The company decided, therefore, that this drive should cease although it was suggested that a four inch bore hole should be sunk down onto the No.3 Level from the 1100ft Level in a last effort to locate the vein and aerate the forehead. The search for barytes in the 650ft and 900ft levels would continue as planned. In the meantime it was decided to evaluate the zinc reserves in No.1 Level - but No.2 Level, now filled with the waste from the driving of the La Porte Incline, would never be used or worked again. The reopening of the No.1 Level proceeded and ore samples were taken, the resulting analysis estimating ore reserves of 40,000 tons at 10% zinc. A reasonable barytes string was also noted.

To assist with the extra power now required in the mine the company obtained two Arpic compressors, each producing 350 cubic feet per minute of compressed air, possibly on indefinite loan from the Caldbeck mines as the ore there was nearly finished. Several more of Caldbeck's miners were in the process of being transferred to Force Crag to assist their mates who had been transferred earlier. Two of these men, Trevor Jones and Harold Collinson, were set to work clearing and retimbering the No.1 Level.

The Big Rise was completed by the end of 1963 without any problems and this meant gear could be hauled straight up from No.3 Level and, more importantly, ore could be sent down. Another important development which also took place around this time was the completion of the rise at the end of the 1100ft Level which had come up into the 80ft Level. Robbie Hilland was on the team which broke through, and no doubt it was a jubilant band of men who descended the La Porte Incline that night. This latter development was an important part of the plan as the miners now had a proper escape route via the High Force workings; the rise also ventilated this section of the mine efficiently, as it still does today. Alongside the miners' ladderway in the new rise were installed twelve-inch diameter pipes which transferred all the water from the 80ft Level down to the 1100ft Level, where it was channelled to the top of the La Porte Incline. Here, as can be imagined, it helped immensely in the process of slushing the ore down to the top of the Big Rise. A telephone communication system was also set up so that the winchmen were always aware of when ore was tipped into the incline and needed to be slushed down.

Early in 1964 Robbie Hilland and Ronnie Coulthard went to work the High Force Crosscut and commenced mining the stopes. No more ore, from this time onwards, was to come out of the Crosscut but was sent instead down the recently completed ore-passes to No.3 Level. The 650ft Level was pushed forwards and after 200ft a very good show of barytes was cut. The level was driven a further 200ft with the vein in evidence. At last it seemed that things were looking better. At the end of the level a ladder rise and ore-pass was blasted to the 900ft Level, 75ft above. Stan Mawson, of Caldbeck, was one of the best rise men in the district and he, with his mate Tom Nicholson, was given this task. By January 1965 they had made the connection with the 900ft Level.

The ore discovered in the 650ft Level was now prepared for stoping - things were going well. Stan and Tom were immediately engaged in putting up the final rise from the 900ft Level to the 1100ft Level, thus completing the series. This latest rise caused problems from the start, the rock, after going up a few feet, turning to rubbishy clay and water cascading onto the men. Eventually the rise broke through into the 1100ft Level; it was named "Mawson's Rise" in tribute to the man who had spent weeks working in the deluge of freezing water and had stuck to his craft in spite of the adverse conditions. Meanwhile, ore was being sent down from the upper workings by Robbie Hilland, Ronnie Coulthard and Bob Young. Quite a lot was being won from the 50ft Level and some from a good pocket recently discovered just below the 80ft Level, which was estimated to contain 3,000 tons.

Production of ore for 1964 was 402 tons of dressed barytes with a further 924 tons in 1965. On completion of the last rise, and with ore ready in the stopes, it was hoped that the mill would be capable of producing 80 tons of dressed barytes per month. Caldbeck mines were now shedding more men and the following came to Force Crag during this period: W. Attigen, S. Cartmell, R. Young, and J.E. Hewer who later became mine foreman. The total workforce at the mine was now six miners, a slusherman, loco driver, lorry man, fitter, three millmen, and a foreman-manager. The wage bill for the mine was around £180 per week.

The planned budget had been exceeded by £11,500, of which £4,000 could be directly attributed to work on the rises. Also, much expenditure was due to the hard nature of the ground making drilling expensive and increasing gelignite costs. When rotten ground was encountered it had all to be timbered, and again this ate into the budget. But now all the preparatory work had been completed and the ore was about to be taken. This, however, was not to be immediate and for some inexplicable reason, when the mine was ready to go into full production, a policy of further exploration was adopted. Ore was still being produced to keep the mill running but at nowhere near its capacity. Attention was turned again to the Zero Level and No.1 Level, neither of which had been looked at since the 1920s. Consideration was given to pumping out the old sump in No.1 and determining the strength of the lode. Diamond drilling was also considered, some core samples being taken but then little else happened.

Barytes was being brought from the workings at a steady rate, the new series of rises delivering the ore down to No.3 Level efficiently enough and ore, still being won from the upper stopes and tipped on the incline, being slushed down to the Big

Willie Shaw (centre) and his men about to commence driving the No.3 Level from the foot of the La Porte Incline, 1963. (Hilland)

Robbie Hilland and Owen Beattie driving the Big Rise, 1963 (Hodgkins)

The No.1 Level portal in the 1960s. (Hodgkins)

Battery loco on the No.1 Level tip. (Hodgkins)

Rails from the 900ft Level crossing the La Porte Incline on a bridge of wooden sleepers. Ore from the hoppers on the 900ft Level was tipped off the bridge onto the sole of the incline to be ''slushed'' down to the head of the Big Rise. (Findlay)

Landing stages and slusher winch station at the top of the Big Rise on the La Porte Incline. (Tyler)

Slusher (immediate foreground) and slusher winch at the head of the Big Rise on the La Porte Incline. A paraffin storm lantern still hangs on the wall, a memento of busier times. (Tyler)

Looking down the La Porte Incline with the rails of the 900ft Level leading to the foot of Mawson's Rise. (Brown)

An old winch rusting in the shadows of the 1100ft Level. (Tyler)

An Ore hopper on the 1100ft Level at the top of the La Porte Incline. (Brown)

An iron chain ladder disappearing into the darkness above the 100ft Level at the top of the la Porte Incline. (Tyler)

On the 1100ft Level - wooden ladders and water pipe rising into the darkness and the workings of the High Force 80ft Level. (Tyler)

On instructions from MESSRS. McKECKNIE CHEMICALS Ltd.

FORCE CRAG MINE, BRAITHWAITE, KESWICK

THURSDAY, 16th MARCH, 1967

PENRITH FARMERS' & KIDDS'

will sell by Auction on the premises
the following

Mining Plant . Equipment . Machinery and Stores

Sale to commence at 11 a.m.

Break for Lunch, 12-30 to 1 p.m.

VIEW : Wednesday, 15th March, 10 a.m. to 4 p.m.

TERMS : Strictly Cash.

Catalogues, price 1/- each, obtainable from the Auctioneers,
3, Lake Road, Keswick.

N.B. The quantities as stated are believed to be correct but no guarantee is given.

Rise with reasonable ease. But there seemed to be little interest among the miners. The men from Caldbeck were disenchanted with their lot, and bonuses were poor in comparison with their earnings at Potts Ghyll.

By 1966 the ore situation in the High Force was critical, many of the estimates given as reserves having taken into account the ore pillars which were actually supporting the stopes. In the old days these would have been taken out to retrieve the last few hundred tons, and might even have been replaced with timber for support, but here there was a dilemma. If the pillars were taken out the mine would become unsafe and should a collapse occur this would prevent the miners having a secondary exit. This situation would effectively close the mine for ever. With regard to timbering, this would be a costly exercise and in most cases the stopes were so wide as to render this totally impracticable. In view of this situation, then, the remaining barytes was taken from the stopes, leaving the pillars untouched. From the High Force Crosscut a small amount was removed which had recently been discovered by Robbie Hilland and his team, who had been doing some final exploration work. It was decided that the mine would not be robbed and it is as a result of this decision that the mine still stands today.

John Simpson, the mine manager, was a hard working fellow and grafted alongside his men to ensure the mine was worked to the best advantage. This they tried hard to do, for a lot of money had been put into the venture by McKechnie Brothers and a fantastic amount of barren driving done in order to prove the vein. In reality, though, the vein had not been located at depth, despite the discovery in the small stope in the No.3 Level. The Company was not prepared to invest in another drive off the La Porte Incline below the 650ft Level where, supposedly, the vein must lie.

The production figures speak for themselves, the largest production having been only 371 tons in a month, and the heaviest single lorry load to leave the mill having been just in excess of 19 tons. The overall production tonnage was somewhere in the region of 15,000 tons of ore milled, during the McKechnie tenure, which averaged at just over 2,000 tons per year. With rising costs inevitable, the prospect of finding the elusive vein was too much of a risk and so McKechnies closed the mine. On the 16th of March 1967 the plant was auctioned by Penrith Farmers & Kidds. Of the 217 lots on offer, most were sold and the total raised was £2,917.12.

6. A FUTURE WITH WILLIE SHAW

The mine did not stay idle for long for Mr. W.T. Shaw, the Keswick mining engineer who had been on McKechnie's payroll, felt confident that Force Crag Mine had a good future. It was because of his local knowledge and contacts that a new company was formed, Force Crag Mines Toronto Limited, based at Suite 313, 67 George Street, Toronto, Canada. Apparently a share capital of $4,000,000 was floated by the consortium. Work commenced at the mine, for the retrieval of lead and zinc, on the 5th of June 1967. Many of the men of the old workforce returned when they realised the mine was to reopen; among them were Owen Beattie, Robbie Hilland, William Hodgson, David Johnstone, T. Nicholson, J.W. Simpson, Alf Gaskell, and of course Mr. Willie Shaw. The men were started on £10.25 per week with a bonus of a £1 per foot when driving rises and levels. The day started at 8:00am and normally finished at 3:00pm, Monday to Friday, blasting always being done last thing in the afternoon. Amazing though it may seem, electric lighting had still not been installed at the mine, even with the Canadian influence. Underground, carbide light was still the order of the day, the two main types of lamp used being the Premier King No.61A, which could burn for ten hours and give a twenty-candle power light, and the Wolf 854 with three-and-a-half inch reflectors. The latter burnt for only eight hours with a candle power of sixteen, and cost £1.75 with carbide at £8 for 100kg. Carbide lamps give off a very warm and white light and have been the miners' friend for nearly a hundred years.

The first job the company embarked upon was the driving of a new entrance to the No.1 Level, for although only recently repaired the weight of the overburden was so great that without constant attention the timberwork was prone to collapse. The new No.1 was driven through the scree for 90ft using timber from the local forests of Swinside and Thornthwaite. When the level was secured as far as the solid rock it was then blasted on a further 80ft to join the original level. At the junction can be seen the difference in size, the old level being much smaller, the new level being cut to 7ft 6ins high and around 5ft wide. This work completed, some of the men were set to work on Zero, the "Unsuccessful Level". This was generally regarded as a

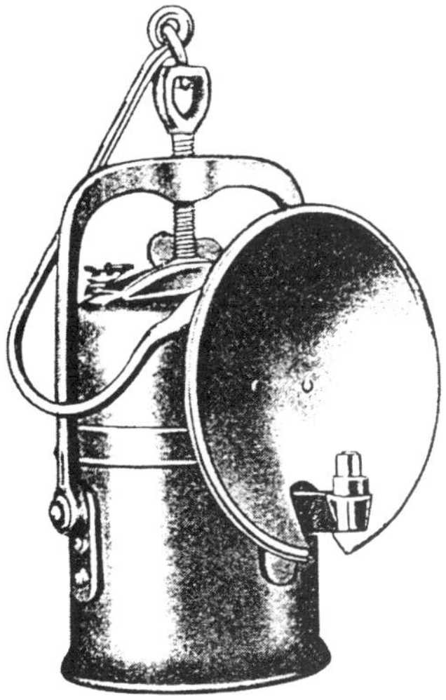

King Lamp. No. 61.

MINING or ENGINEERING LAMP

with

POSITIVE SCREW LOCK

In Steel (tinned after manufacture) with
Brass Water Vessel Base

No. 1 size is also supplied in Brass

Three sizes	No. 0	No. 1	No. 2
Height to Bridle ...	$9\frac{3}{4}$ in.	$8\frac{1}{2}$ in.	$7\frac{1}{2}$ in.
Weight	3 lb.	2 lb. 6 oz.	1 lb. 12 oz.
Charge of Carbide ...	16 oz.	10 oz.	6 oz.
Lighting Power ...	25 c.p.	20 c.p.	16 c.p.
Burning Capacity ...	10-12 hrs.	10-12 hrs.	8 hrs.
Burner No. 105 ...	21 litre	14 litre	14 litre
Code Word	AKIG	AKIN	AKIL

Alternate Hooks can be fitted to either Lamp
Reflectors A B and D (page 11) are interchangeable

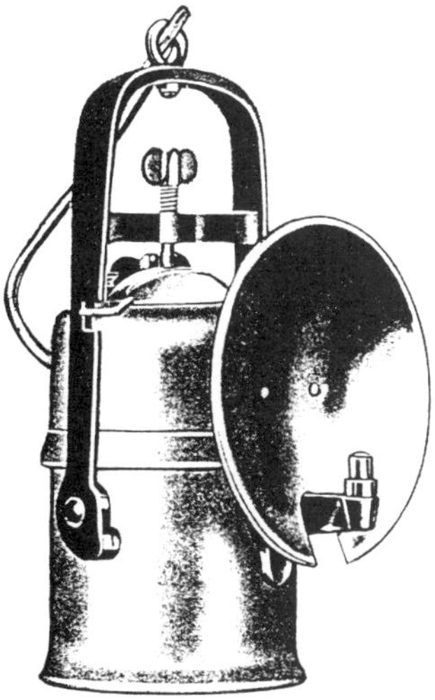

Cumberland Lamp. No. 73.

MINING or ENGINEERING LAMP

with

POSITIVE SCREW LOCK

STIRRUP HANDLE WITH LEVER TIPS GIVING
AUTOMATIC RELEASE OF SEALING JOINT

In Steel (tinned after manufacture) with
Brass Water Vessel Base

One size in Steel	
Height to Bridle	10 in.
Weight	3 lb. 0 oz.
Charge of Carbide	10 oz.
Lighting Power	20 c.p.
Burning Capacity	10-12 hrs.
Burner No. 105	14 litre
Code Word	COLO

Alternate Hooks can be fitted to Lamp
Reflectors A B and D (page 11) are interchangeable

waste of time, the timbers being in a dreadful state, but out of curiosity more than genuine mining interest the level was cleared to see what lay beyond. Some people say the original 100 yard drive cut the vein, while many dispute this claim; but if the drive did cut the vein, why, then, had it not been exploited at the time? The new company decided to drive off west through the shale and after progressing 120ft, through unstable rubbish, solid rock was reached. Robbie Hilland, Owen Beattie, David Johnstone and Trevor Jones were some of the men who pushed this level and were well pleased to come into solid ground. The company's plan was to push the Zero Level under the No.1 to see the vein at depth, while at the same time the No.1 was also being extended. The "Old Men's Sump", sunk in the sole of the No.1 Level during the 1920s, was to be drained and sunk the remaining 80ft or so down to Zero Level to ventilate the mine and provide an escape route for both levels.

The levels were driven forward with the best equipment money could buy, the face drilling being done with Atlas Copco boring machines and Sandvik drill steels. In view of the expense of the drill steels, which were tungsten-carbide tipped, all had to be accounted for to the manager. During the first year 1,082lbs of gelignite was used, costing the firm £108.68. Fuses cost £49 per 1,000 and were available in two varieties - capped fuses for use mainly on driving levels and blasting small rises, and electric detonators for rises over 60ft and for use in very wet conditions. It appears that little or no ore was mined during this initial period but the ore-pass and ladderway rise, from No.1 to Zero Level, was completed in 1970. Completion came as a relief to the men who had been cutting the lower level, as conditions had become most unpleasant at the face. The miners driving in No.1 cut a spring and this level was making a lot of water, most of which cascaded down the new manway. However, a new problem was about to strike the mine: vandalism. The men, on arriving one morning, found the office wrecked and windows smashed in the mill. Regrettably, this nuisance was to recur; on one occasion the vandals managed to get underground and cause more damage. This had to be stopped - so a night watchman was engaged.

As part of the company's expansion programme a special feasibility study was carried out in 1971 to prove the veins south-east of the sett but only two small lead veins were found. These were of no significance and were totally impracticable to work.

As driving progressed in No.1 Level a pocket of radon gas was unexpectedly released from the rock and, as a precaution, the level was evacuated then walled up some 600ft from the face. A two inch diameter pipe was fixed into the base of the wall to allow the copious amounts of water to escape. Just back from the dam is the final phase of the company's preparatory work. Shaw's Rise, a second ore-pass and manway, was put up from Zero Level directly below the old No.4 Stope, which in the old days had been the site of Horn's Rise. Whether or not it was the intention of the company to rive out the old deads which were still lying in the stope, and extract their mineral content, will never be known. Zero Level, at this point, having been driven forwards 1,800ft purely as a development drive, shows very little vein stuff with no stoping done. In Shaw's Rise, 50ft below No.1 Level, is a small crosscut but no other working. The rise was put up by Tom Nicholson and, again, the size is of ample proportions. Alas, Tom's good friend from Caldbeck, Stan Mawson, had now retired because of ill health. Other men working at the mine at this time were J. Lavery, J. Shepherd, W. Kelly, N. Shepherd, and H. Graves. Sadly, on the 20th of June 1972, J.W. Simpson, the mine manager, collapsed and died at the mine. This was a sad blow as many of these men had worked together for all their working lives in the various mines of the district.

Another feasibility report was done and the results this time promised a total reserve of 48,500 tons of ore in Zero and No.1 levels. The ore showed 2.33% lead and 4.0% zinc, but little work was being done and the workforce was informed that the mine would close on the 18th of August 1972. The law states that when a mine is closed all levels must be secured, shafts capped or fenced, and open stopes secured in a similar fashion before the mine is abandoned. A team of men was busy doing just that, on the 10th of August, at the open stopes high on the fell, which offered a vertical drop of some 280ft to the High Force Level below. One of the miners, intrigued by his strange surroundings - for he hadn't been involved in working in the upper mine area - decided he would take a walk and explore inside the High Force Crosscut. Unfortunately for him, sections of the floor are under-stoped and, without warning, the area of timber floor on which he was standing collapsed and he fell some 50ft to a sub level. His workmates quickly raised the alarm and he was rescued by the Keswick Mountain Rescue Team, suffering minor injuries, a very lucky man indeed. The news hit the local headlines and the event was even recorded in the national press, so Force Crag Mine was, on this occasion, closed in a blaze of publicity.

The mine once again fell into neglect and was soon the target for every vandal who visited the Coledale valley. Windows were smashed and doors ripped off and soon the buildings became the habitat for dozens of herdwick sheep. Up in the High Force a gentleman of the road took to residing in the old miners' hut. He made a heather bed on the floor and used wood which was lying around to burn in the stove for warmth and cooking. He adorned the walls with 1960s style hippy flower-power art, though he did not stay long and his faded wall paintings are now to be seen only faintly.

It was not until 1977 that details of a new lease were finalised. This new operation was to be called Force Crag Mines UK Limited and the directors were

G.W. Hall, M. Klyman and R. Gunn. The intention of this company was to mine the zinc reserves and attain the magic figure of 100,000 tons in order to satisfy their wouldbe backers. If the company could prove the reserves, a large capital injection would be available. The active workforce included Robbie Hilland, Paul Ratcliffe, George Hughes, Kenny Hill, Clive Duke, Maurice "Moss" Dunkley, Walter Muir, Owen Beattie, and Robert Gunn who was also a director. It was decided that the most prudent objective was to prove the vein at depth, below the Zero Level, something never before considered by the previous companies. Robert Gunn was a skilled man and realised that sinking a shaft in the sole of the level, to a depth of 30 metres, should, if the vein was found, give him the quantity of reserves required to secure the investment.

The mill was cleared of its recent inhabitants and copious amounts of disinfectant applied. No.1 Level entrance was again in a sorry state so a new portal was erected, consisting of breeze blocks and railway sleepers. The site chosen for the new shaft was at the end of a short heading 750ft from the entrance of Zero Level. Shaft sinking was a tough job, all the waste having to be mucked out by hand to start with - back breaking work with water constantly pouring into the hole and making conditions appalling. Robbie Hilland was one of the miners drilling the shaft, using the Holman "sinkers" drilling machine. He would bore up to forty 4ft holes per round; the charges would then be set and fired, and then mucking out would begin. Obviously, with each set of rounds fired, the distance for hauling became greater. The shaft was set up with a powerful winch, operated by Walter Muir who ensured that the contents of the kibbles were tipped in a large receiving hopper which had been constructed at the shaft collar. Bob Gunn realised that cleaning out by hand was strenuous and time-consuming, so he designed a unique hydraulic shovel system. Basically, after each blasting, a JCB arm and bucket on hydraulic legs was lowered into the shaft. The legs were rammed into the sides of the shaft and as soon as they were secured the JCB bucket mucked out into the waiting kibbles. The hopper was emptied into ore tubs and the waste hauled out to day by the electric loco driven by Owen Beattie. Owen was also required to operate the Eimco 12B, a type of face-loading shovel which is capable of loading tubs.

When the shaft was down over 20ft the men installed a pump; George Hughes used to go down during the night or early morning to ensure that the shaft was clear of water before the miners arrived. Day after day, in this relentless manner, the shaft was sunk to the required depth of 30 metres. From the foot a crosscut was driven which cut an eighteen inch vein of barytes and zinc. The vein was then followed west for a distance of 250ft to a point where it nipped out to virtually nothing. This was just about the worst thing that could have happened and the company, realising the gamble had failed, was not prepared to put any money into the venture. The mine closed virtually straight away, in 1978. The shaft soon flooded, its depths now a mystery to all but those who worked there.

Bob Gunn was a tenacious man and was convinced, as Willie Shaw had been, that the mine could be worked successfully. He struggled to raise capital and, eventually, was able to hit his target. He managed to negotiate a sub-lease and, on completion of the formalities, formed the Braithwaite Mining Company Limited in 1978. Once again the local miners were quick to respond to the challenge and most returned to work with Robbie Hilland as foreman.

The company policy was to inject sufficient cash to provide a small but efficient mill, refine as much of the existing broken ore, lying in No.1 Stope, as possible and also any other ore found that was ready for milling. If the mill could be put in order, and all the readily available ore milled, this would create some revenue to help support the company. The plan seemed basically sound but for some reason the mine was redeveloped and the ore waiting to be milled was left untouched. The fitting out of the mill went ahead, during which time an inclined shaft was commenced a short distance inside the entrance of No.1 Level, this being driven to strike the vein under the old No.1 Stope. The shaft was cut in three distinct legs, the first going down about 30ft and in a northerly direction, where on reaching the vein it was continued west. After a short distance the vein widened to show 4ft of quality barytes, which was removed with pick and shovel though, when available, an ordinary pneumatic drill known affectionately as a "windy pick" was brought in for assistance. The drive continued down for another 40ft at which point a vertical 40ft ore-pass was sunk down to Zero Level. Soon after a manway was driven parallel to the ore-pass to connect with the working, so that the miners could gain easy access. The men who drove the rise and ore-pass were Clive Duke and "Moss" Dunkley.

The mill, apparently, was completed at way above the anticipated cost - but still ore was not fed through. More exploration was carried out, this time further along the No.1 Level, adjacent to the old No.2 Stope, at the point where the old men had sunk their sump. Here, an inclined rise was driven a distance of 200ft, ending in a vertical 30ft section which holed through into the sole of the old No.2 Level. Off the incline were driven six very short drives heading north and south. This inclined rise, now known as Gunn's Rise, was driven by Clive Duke and "Moss" Dunkley. The incline was very difficult to work as the No.2 Level was filled with thousands of gallons of water, trapped in the old stopes. This water, over the years, had permeated the fractured rock, making life most unpleasant for the miners working below. Even today, a walk up Gunn's Rise is, to say the least, an invigorating experience. One can only speculate why this expensive exploration was carried out when the mill was ready and money was tight. Maybe it was the expectation of finding the old stopes of the No.2 Level bulging with ore; maybe it was the possibility of larger

Owen Beattie filling tubs from a hopper on Zero Level. Carbide lamps - the miners' friend - hang from a joist in the roof. (Hilland)

Willie Shaw (far right) and his men at the Zero Level portal, 1968. (Hilland)

The High Force. An erstwhile resident, a wandering hippy sits at the door of the bait cabin he decorated with flower-power murals. In the distance the old ropeway pylons stand rusting, prior to their removal. (Davies-Shiel)

Top of a ladder rise above the 1922 sump on the No.1 Level. (Tyler)

Receiving hopper straddling Gunn's flooded shaft on the Zero Level. (Tyler)

Gunn's modernised No.1 Level portal. (Tyler)

The kibble used in the sinking of Gunn's Sump stands outside the Zero Level, 1980. (Tyler)

FOR SALE BY AUCTION
ON
THURSDAY, 1st JULY, 1982
AT 1 p.m.
PLANT, EQUIPMENT, ENGINES, VEHICLES AND TOOLS
AT
BRAITHWAITE MINES, BRAITHWAITE, KESWICK, CUMBRIA.

Briefly comprising: Motors and gearboxes, control panel, idlers and drums for conveyor, bench and vice, grinder, Coote vibrating feeder and controller, 2in. mono pump, engineers tools, 35 ton steel ore hopper, apron feeder, Baxter 20 x 12 jaw crusher, Jenkins 'Viking' vibrating screen ⅛ mesh, 18in. x 24in. belt conveyors, 2 Wemco 66 flotation cells with 10 HP agitators, 6 flotation cells with 5 HP agitators, 12ft. x 6ft. conditioning tank, Babcock and Wilcox ball mill with Perkins diesel engine, belt feeder, Regent pump, Nash vacuum pump with 25 HP motor, 2 Dewer disc filters, starters and isolators, 60 metres of soil pipe, 650 gallon oil tank, 1,000 gallon water tank, 50 metres 3in. plastic pipe, International rough terrain fork lift truck, Ford Transit van, 250 amp welder, scrap and many other lots.

VIEWING: Morning of sale from 11 a.m. onwards.

All accounts to be settled on day of sale.

reserves being proved. If this was the case, armed with this information, more money could be raised for the final kill.

Alas, fate was to be more than cruel for 1981 brought one of the Lake District's notoriously bad winters, the freezing conditions running right through to spring. A few tons of ore were eventually run through the mill but, apparently, the percentage of mineral recovered did not come up to the original assay. This shortfall was put down to the fact that, over the years, the stock-piles had been well picked over by people fossicking and this had reduced the mineral content. Dejected and short of cash, the men persevered until one day in 1982, on arriving for work and entering the mine, it was discovered that a large section of No.1 Level was in a state of collapse. This was the final straw for despite hundreds of tons of scree being cleared by mechanical diggers - to avert the pressure on the straining timber in the level below - they realised they were fighting a losing battle. This was a bitter blow for everything was ready, with zinc fetching £420 per ton and lead £320. But there was no more credit and the bank foreclosed. The mine was shut down and a sale of all the plant and equipment held on the 1st of July 1982. The last remaining miners, Robbie Hilland, Owen Beattie and a few others, had lost their place of work, some of them having been employed at Force Crag for over twenty years. Now it was gone, and so had all but one of the Lake District's metal mines.

No. 854

This is a new lamp of a size between that of the **Star** and the **Norton** and combining all the advantages of these types with extended simplification of working parts.

No.	Weight lb. oz.	Height inches	Burning Hours	Candle Power	Carbide oz.	PRICE Steel	Brass
854 ..	2 7	8	7 to 8	16	7	9/-	—

7. THE NEW COLEDALE MINING CO.

Nearly two years went by before a lease was taken out on the mine, all the legalities being completed and signed by May 1984. The mine was leased by four people, namely Mr. Peter Blezard, Mr. Mike Sutcliffe - a fine engineer both topside and underground - Mr. Lindsay Greenbank and Ms. Anne Danson. The company is called the New Coledale Mining Company, and Force Crag is now the last working mine in the Lake District. The owners are faced with a race against time as the mine is on a short lease and they are aware that it will be forced to close when the lease expires (with little or no chance of renewal) in the early 1990s. The ores currently being extracted, from the low levels No.1 and Zero, are zinc, lead and barytes. Besides the four directors, a regular team of part-time miners assisting at the mine includes Brian Stevenson, Ronnie Calvin, Neil Barnes and Dave Blundell.

The first task confronting the new owners was to clear out the mill and install the machinery, a long and arduous task which tested the skill and ingenuity of the workforce. However, major problems appear to have been solved and the mill is now working. During the interim setting up period other modifications were done, including total renovation of the Ball Mill, adjustments to the Davidson's Jigs, and many more minor works. The weather has, of course, brought its own problems. During 1988 a 600 gallon fuel tank was torn from its mountings and blown fifty yards down the road, while the gales of February 1990 ripped off part of the mill roof and blew a four inch diameter water pipe from its anchorage and snapped it in two. Vandals have, regrettably, visited the mine - which is a constant niggle - but a watchful eye is kept on things and the mine is regularly patrolled by the National Park Warden.

Nothing much has changed in 150 years of mining at Force Crag. While the mill was being put in order the levels were, simultaneously, being prepared for the extraction of the ore - zinc being the main target. The entrances to both No.1 and Zero levels were partially retimbered and made sound, and the decision taken to develop a pocket of ore in Zero Level, virtually underneath the old No.4 Stope in No.1 Level. Here, a rise was put up 20ft and a small sub level driven off to explore

The rocky face of Force Crag towering over mine and mill. (Tyler)

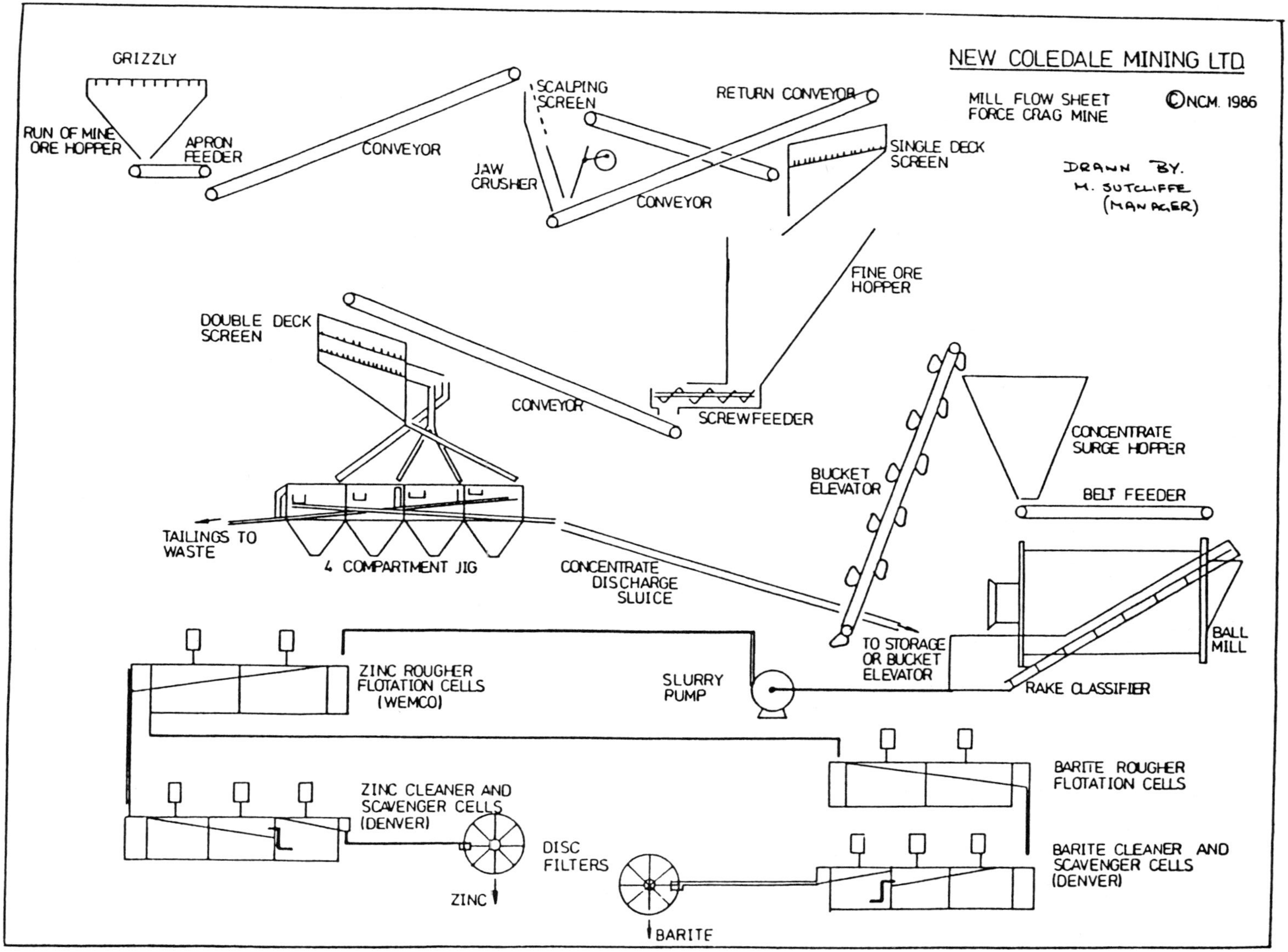
NEW COLEDALE MINING LTD
MILL FLOW SHEET
FORCE CRAG MINE
©NCM. 1986
DRAWN BY.
M. SUTCLIFFE
(MANAGER)
GRIZZLY
RUN OF MINE ORE HOPPER
APRON FEEDER
CONVEYOR
SCALPING SCREEN
JAW CRUSHER
RETURN CONVEYOR
CONVEYOR
SINGLE DECK SCREEN
FINE ORE HOPPER
DOUBLE DECK SCREEN
CONVEYOR
SCREWFEEDER
TAILINGS TO WASTE
4 COMPARTMENT JIG
CONCENTRATE DISCHARGE SLUICE
BUCKET ELEVATOR
CONCENTRATE SURGE HOPPER
BELT FEEDER
TO STORAGE OR BUCKET ELEVATOR
BALL MILL
RAKE CLASSIFIER
SLURRY PUMP
ZINC ROUGHER FLOTATION CELLS (WEMCO)
ZINC CLEANER AND SCAVENGER CELLS (DENVER)
DISC FILTERS
ZINC
BARITE
BARITE ROUGHER FLOTATION CELLS
BARITE CLEANER AND SCAVENGER CELLS (DENVER)

the vein, which shows good barytes and lead, and also some fluorspar. A parallel rise was blasted to connect with the sub level and a wooden hopper constructed at its foot (it was during the driving of this sub level that I personally experienced the delights of being present while a series of charges went off. The holes were drilled, the charges laid, and we retired discreetly to a short crosscut some forty yards down Zero Level. The charges were to be detonated electronically. We sat there in the dark, in anticipation, hands over our ears, when the ground shook and a blast of dust blew by the end of the crosscut. Then, upon removing our hands from our ears, we were greeted by the loudest noise I have ever heard, and again the dust blew past and the ground trembled with a frightening ferocity. Ronnie Calvin, standing nearby, grinning, said in his best West Cumbrian: "Well, Ian, didsta like that?" This working is now called Calvin's Stope, or the 1500 Stope, it being 1,500ft from the entrance of Zero Level).

More exploration work was carried out, this time from No.1 Level, a sub level being driven by Brian Stevenson and Lindsay Greenbank in a good pocket of sphalerite 30ft down Gunn's Incline. A slusher was installed and the ore dragged down the incline into a hopper on Zero Level. During this period Mike Sutcliffe and Peter Blezard were working feverishly in the mill to ensure that the machinery would be ready to receive the first batch of ore.

The years moved on and by 1987 some trial runs had been attempted in the mill but the ore separation process was still not quite right. A specialist was called in and his expertise, together with Mike and Peter's knowledge of the system, produced the desired formula and the mill is at last operational.

During 1989 a third area of ground was explored at the Old Ladderway Rise, now known as the "1200", between No.1 and Zero levels. The miners drove a 40ft sub level into very good zinc and again a slusher has been set up and a quantity of ore drawn off. The mine is now operating two battery locos, capable of pulling six tubs each, one stationed in Zero Level, the other in No.1 Level. Further development work has been carried out in No.1 Level with heavy-duty rails being laid from the receiving hopper at the mill through to the old No.1 Stope, which was last worked with any vigour during the 1920s. Here the zinc ore lies ready broken in the stope and very little work is required to move it. Certainly, the Eimco face loading shovel will cope admirably, for when this little monster rattles into action it is capable of loading the tubs at the rate of twenty men per hour. When this ore has been removed and processed, the zinc concentrates produced will be despatched to the Rio Tinto Zinc Company smelter at Avonmouth. Let us hope that this is the first of many loads and that the last surviving mineral mine in the Lake District has a prosperous future.

Ronnie Calvin boring shotholes in the 1500ft Stope, Zero Level. (Calvin)

Peter Calvin charging shotholes with Special Gelatine and electric detonators, 1500ft Stope in Zero Level. (Calvin)

On the bank outside the No.1 Level. Ronnie Calvin and Lindsay Greenbank with a tub of ore for the crusher. (Tyler)

Lindsay Greenbank hauling tubs from the No.1 Level. (Tyler)

Timberwork in the entrance of Zero Level. (Calvin)

Peter Calvin filling tubs from the 1500ft Stope hopper on Zero Level. (Calvin)

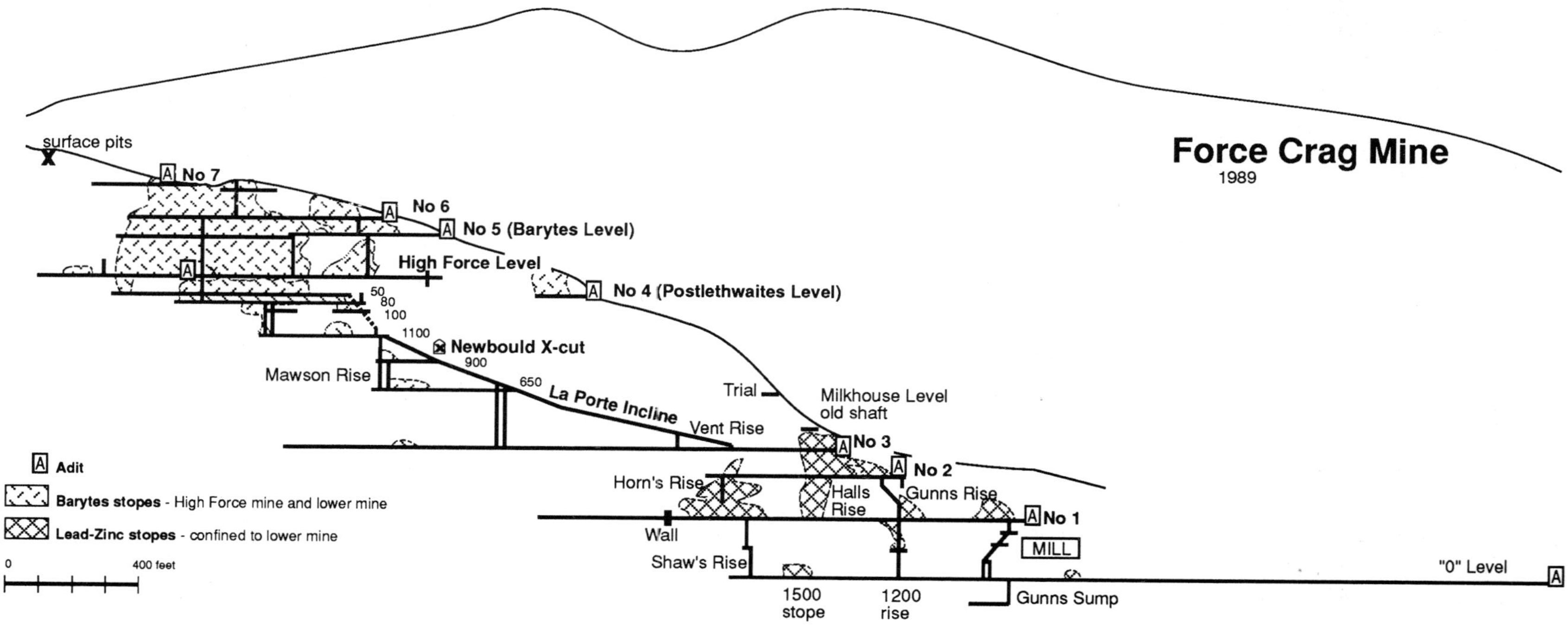
Force Crag Mine
1989
surface pits
No 7
No 6
No 5 (Barytes Level)
High Force Level
No 4 (Postlethwaites Level)
50
80
100
1100
Newbould X-cut
900
650
Mawson Rise
La Porte Incline
Vent Rise
Trial
Milkhouse Level
old shaft
No 3
No 2
Horn's Rise
Halls
Rise
Gunns Rise
No 1
Wall
MILL
Shaw's Rise
"0" Level
1500
stope
1200
rise
Gunns Sump
Adit
Barytes stopes - High Force mine and lower mine
Lead-Zinc stopes - confined to lower mine
0
400 feet

Force Crag Mine, as it stands today, has been working close on 152 years and during this time has produced some 60,000 tons of barytes, 1,248 tons of zinc and 624 tons of lead, this last containing many ounces of silver. Something like 5,500 yards of main level have been driven, much of this by hand, and 1,840 feet of shafts and rises. It has provided many men with employment over the years and has brought success to some, and to others financial ruin. All these men, though, had certain qualities in common: the spirit of adventure and the will and the courage to work against nature in this strange underground environment, where there are no shadows and the temperature hovers around fifty-five degrees Fahrenheit.

I am glad to say that Force Crag is a friendly mine. During the course of writing this book I have, with others, gained access to areas of the mine which have been sealed off for many years, some areas, indeed, not having been visited since that section of the mine closed. Many artifacts have been retrieved and are now on display in the Caldbeck Mining Museum. The variety of these relics is extensive, one superb example being a miner's water bottle discovered in the old No.2 Stope in No.1 Level in 1989. Access to the No.4 Stope on No.1 Level was gained by climbing a marvellous cobble-lined rise built through stacked deads. At the top of the rise is an enormous stope some 250ft long, 130ft high, and 12ft wide. Still bolted to the walls are the ladders put in by the old men in Horn's Rise, which disappears into the elusive No.2 Level. This section was, without doubt, our greatest discovery.

Another interesting area of the mine is located above No.2 Stope in No.1 Level. After climbing Gunn's Rise we scaled the final 30ft in appallingly wet conditions to gain access to No.2 Level, and although we were not the first to visit this area we were certainly the first to gain access to the higher stopes and the top of Hall's Rise and make our exit by No.3 Level.

The whole of the La Porte Incline has been explored and the 200ft Rise, the Big Rise, descended on abseiling ropes, just in case there were any levels driven off. In the High Force Mine, on one very cold day in 1982, we entered the No.3 Level and proceeded up the La Porte Incline and onto the 1100ft Level which, at that time, was virtually chest-deep in water. We explored the level as far as the rise and ladderway (up to the 80ft Level) which is still intact and where many artifacts were abandoned. The level has been subsequently drained and does not now offer such a sporting route to the 80ft Level. I can well remember the horrendous climb up the ladderway from the 80ft Level to the workings on the High Force Crosscut, when the water simply poured onto us as we progressed onwards and upwards. Much restoration work has been carried out in the High Force Crosscut, which suffered a major collapse around 1987 at the point where the adit level intersects the vein. In 1989 four wooden Hudson's mine trucks were dug out of the collapse and mounted on rails - and well they look.

The 50ft Level was a source of numerous artifacts, indicating that no one had visited here since its closure. The No.4 Level remains as it was in 1869, with the angled rail still in situ. I believe that our second most important discovery, again in 1989, was the digging of the entrance of No.5 Level. After the initial short drive the whole area has been stoped out over a distance of 200ft, in length, at least 150ft in height, and something like 15ft across with not a single ore pillar remaining. Nothing supports this massive void except two timber manways which rise from the High Force workings, cut through No.5, and continue all the way up to No.6 Level. They are a tremendous sight. The two wooden towers testify to the ability of the miners and joiners who built them. We were unable to scale the ladders, because of their unstable condition, but we were able to descend to the High Force forehead, and east again, places not visited since the High Force closed. Again we retrieved artifacts, among them an early Patterson battery lamp, circa 1945.

During a descent of the gully from near the No.4 Level down to No.3 Level, via the old miners' track, we discovered the pins for the wooden chutes used by the old men for transferring the ore from the upper workings. Also discovered here was a complete bucket unit from the aerial ropeway.

The upper reaches of the mine, No.6 and No.7 levels, are in a serious state of collapse and, I feel, too dangerous for exploration. The secrets which may remain in this part of the mine must stay hidden for the moment. Parts of this area were last worked over 120 years ago, in 1869. I personally attempted to explore this last remaining area but after abseiling into the stopes considered the ground too unstable. I noted, on my descent, the remains of a collapsing wooden manway and a few rotten barytes pillars supporting huge shields of rock. The vein here is not as vertical as in other parts of the mine - and the hanging wall really does hang! I have decided that the mine is entitled to keep some secrets.

Mine exploration is a serious undertaking and should not, under any circumstances, be attempted without the necessary equipment and skills. It should also be noted that Force Crag is a working mine and any approach made to the underground workings should not be done without permission.

TOPOGRAPHY OF FORCE CRAG MINE

NAME	ORDNANCE DATUM	DATE (CIRCA)
Braithwaite Trials	500ft	1870
Coledale Coffin Level	600ft	1564?
Original Mill Site	800ft	1840
Zero Level	841ft	1914
Present Day Mill	900ft	1908
Old Lead Mill	920ft	1854
Old Barytes Mill	925ft	1867
No.1 or Great Level	941ft	1849
No.1 Level (new adit)	939tf	1967
Hand-Dressing Floor	1053ft	1854
No.2 Level	1085ft	1839
No.3 Level	1173ft	1847
Top of Old Shaft or Hall's Rise	1245ft	1870
Milkhouse Level	1265ft	1907
Open Trial on Vein	1350ft	1578
Newbould Crosscut	1550ft	1930
No.4 or Postlethwaite's Level	1644ft	1867
High Force Crosscut	1710ft	1929
Open Stope on No.4 Level	1720ft	1867
No.5 or Barytes Level	1833ft	1869
No.6 Level	1883ft	1873
No.7 Level	2000ft	1940
Trial Pits	2050ft	1869

SOME OF THE MINERS WHO WORKED AT FORCE CRAG MINE

1907	Robert Hindmarch	
1907	James Hindmarch	
1910	William Hindmarch	
1913	I. Davison	
1913	G. Davison	
1914	J. Pearson	
1914	Ned Graves	
1914	D. Horton-Smith	Director?
1914	Mr. Lowe	
1914	Joseph Frost Lobb	
1914	Thomas Crawford Dennison	
1914	Mr. Foster	
1914	G.F. Wallace	Mine Manager
1936	W. Horseley	
1955	G.H. Turner	
1962	R. Hilland	
1962	P. Ratcliffe	
1962	S. Mawson	
1962	R. Young	
1962	R. Coulthard	
1962	T. Nicholson	
1962	W. Attigen	
1962	S. Cartmell	
1962	J.E. Hewer	
1962	O. Beattie	
1962	H. Collinson	
1967	W.T. Shaw	Mine Manager
1967	D. Johnstone	
1967	Willie Hodgson	
1967	J. Lavery	
1967	J. Shepherd	
1967	John Hodgkins	
1967	W. Kelly	
1967	N. Shepherd	
1967	H. Graves	
1967	J.W. Simpson	
1970	A. Gaskell	
1977	G. Hughes	
1977	K. Hill	
1977	C. Duke	
1977	W. Muir	
1977	M. Dunkley	
1981	R.I. Gunn	
1989	Anne Danson	
1989	Peter Blezard	
1989	Mike Sutcliffe	
1989	Lindsay Greenbank	
1989	Neil Barnes	
1989	David Blundell	
1989	Ronnie Calvin R.M.	
1989	Peter Calvin	
1989	Brian Stevenson	

MINERS WHO LIVED IN BRAITHWAITE

1738	Robert Brown	1855	T. Jones	
1796	Anthony Wright	1855	S. Jones	
1813	Joseph Coates	1860	Dixon Barnet	
1817	John Thompson	1861	Samuel Kelly	
1817	Robert Lindsay	1861	Mathewman Barnes	
1819	M. Harker	1861	James Cullen	
1821	Charles Lowden	1862	D. Hardesty	
1832	Thomas Nichols	1862	J. Hardesty	
1843	Thomas Williams	1864	J. Hocking	
1843	Thomas T. Lowden	1864	J. Hocking Snr.	
1844	John Robinson	1867	James Barnes	
1844	James Robinson	1867	Charles Barnes	
1844	E. Martin	1875	John Francis	
1844	William Postlethwaite	1875	W. Francis	Mine Captain
1844	William Postlethwaite Snr.	1876	Joseph Kirkby	
1846	Joseph Horn	1884	Thomas Edwards	
1846	Thomas Horn	1885	Thomas Harvey	
1848	James Rough	1886	Noah Pryce	
1848	Richard Tangye	1886	Samuel Pryce	
1848	Richard Tangye Snr.	1887	William Hughes	
1848	Henry Lowden	1887	James Hughes	
1849	William Hudson	1887	Edward Edwards	
1849	Joseph Hodgson	1887	William Hunter	
1849	Joseph Robinson	1888	* Edward Parry	
1849	George Coulson	1888	John Parry	
1849	Joseph Spark	1888	Joseph Young	
1849	William Spark	1893	Hugh Tonkin	
1849	Henry Osbourne	1894	James Hewer	
1849	Henry Osbourne Snr.	1896	William Clark	
1851	J. Newton	1896	Harrison Hodgson	
1851	James Kelly	1898	* Isaac Grisedale	
1851	William Kelly	1898	John Scott	
1852	Robert Parker	1901	* John Henry Holliday	
1853	William Fisher	1901	* William Holliday	
1853	Thomas McKenny	1904	George Waters	
1854	Thomas Gunwood	1904	G. Harrison	
1855	Jacob Holliday	1906	* Joseph Dixon	
1855	William Holliday	1906	* William Dixon	

1906	* Eldred Lowden	
1907	George Davison	
1907	Alfred Airey	
1908	* William Alexander Kirkby	
1908	* William Kirkby	
1909	* Thomas Edward Turner	
1909	* Hugh Turner	
1910	John Hodgson	
1913	* Nicholas Davies	
1913	* W. Cartmel	
1913	Lancelot Mossop	
1914	* Puetney Porter	Mine Manager
1914	Henry Waugh	
1919	* Herbert Erwin Barnes	
1919	John Barnes	
1919	* Joe Hindmarch	
1921	N. Davies	
1942	* Gerrard Sowerby	
1942	J. Thwaites	Underground Manager
1946	* William Horsley	
1946	* Thomas Dixon Barnes	

* Miners identified as working at Force Crag Mine.

GLOSSARY OF TERMS

ADIT. A horizontal tunnel driven into the hillside for access to the vein or to drain water from the mine.

BACK FILLING. Waste rock which is dumped in a disused area of the mine to save the time and expense of bringing it out to day.

BACKS. Workings or vein directly above the level.

BAIT. Miners lunch or break.

BALL MILL. Revolving steel drum containing steel balls for crushing the ore.

COFFIN LEVEL. A narrow tunnel, coffin-shaped in cross-section, usually dating to the Elizabethan period and driven without the use of explosives.

COUNTRY ROCK. The rock in which the vein is found.

CROSSCUT. A tunnel driven through country rock to cut the vein and often at right angles to the vein.

DAY. Underground term meaning "surface" or "outside".

DEADS. Waste rock stacked neatly to support the mine walls.

DRIFT. Tunnel or level along the vein.

FATHOM. Six feet.

FLOTATION. A system to separate the concentrated mineral by the use of oil rising through water, to which the concentrate has adhered.

FOOTWALL. The underlying wall of a stope or sloping vein.

GANGUE. Worthless minerals, usually quartz and calcite, making up the mineral vein.

GALENA. Sulphide of lead.

GRIZZLEY. A series of iron bars set apart to size the ore.

HANGING WALL. The overlying wall of a stope or sloping vein which tends to form the roof of a working.

HOPPER. A wooden chute fixed into an aperture in the base of a stope so that ore collected above can be discharged into ore tubs in the level below.

HORSE LEVEL. A main adit level where horses are used for underground haulage.

INCLINE. A tunnel driven at an angle, usually between 33 and 50 degrees from the horizontal.

JIGS. Machines for separating mineral and waste by means of agitation and water.

JUMPER. Hand-held drilling steel, various lengths.

KIBBLE. A bucket-like container for hauling ore and equipment up shafts.

LEATS. Courses cut into the fell to channel water for the operation of waterwheels etc.

LEVEL. Horizontal tunnel giving access to workings, also providing drainage.

LODE. Vein carrying mineral.

OLD MEN. Early miners.

ORE PASS. A shaft or chute made of timber, or stone, or cut in solid rock and used for passing ore and waste rock to the level below.

PELTON WHEEL. An enclosed wheel with cups which accept a powerful jet of water. From the drive shaft of the wheel machinery of all sorts can be driven.

PILLAR. A part of the vein which is left in position to support the walls of the mine.

PORTAL. The entrance to a level, usually constructed from local materials.

RISE. An internal shaft put upwards in the vein, which may eventually connect with the level above. If laddered it is then referred to as a manway or winze.

ROYALTY. An area of land determined by a lease, or the payment demanded by the mineral owner on every ton of milled ore produced by the operator.

RUN-IN. When the roof or sides of a level collapse.

SETT. The area of land determined by the lease and on which the operator has freedom to mine.

SHAFT. A hole nearly always vertical but can be inclined, whereby ore is winched from one level to another and by which the miners gain access to the workings.

SLIMES. The fine slurry from the mill created in the crushing process.

SOLE. The floor of a level.

SPOIL. Waste rock discarded outside the mine entrance, usually with no or little mineral content.

STAMPS. A series of large iron hammers for breaking up the ore, usually driven by a waterwheel.

STOPE. An area where the vein has been removed - the working part of the mine.

SUMP. A shaft sunk in the sole of the level.

TAKE-NOTE. A licence to search for minerals.

VEIN. Mineral body, usually of a vertical or near-vertical nature, striking through the rock for great distances and to great depths.

BIBLIOGRAPHY

Adams, J. - Mines of the Lake District Fells, 1989.
Bowen, D. - Mines and Quarries Acts, 1923.
Burt, R., Waite, T., Atkinson, M., Burnley, R., - The Cumberland Mineral Statistics, 1982.
Camm, F.J. - Dictionary of Metals and Alloys, 1941.
Collingwood, W.G. - Elizabethan Keswick, 1912.
Donald, M.B. - Elizabethan Copper, 1989.
Dunham, K.C. - Barium Minerals in England and Wales, 1945.
Eastwood, T. - Memoirs of a Geological Survey Vol XXII, The Lead and Zinc Ores of the Lake District, 1986.
Landless, V. - Force Crag Today, 1976.
Lingenfelter, R.E. - The Hardrock Miners, 1974.
Postlethwaite, J. - Mines and Mining in the English Lake District, 1913.
Richards and Locks - Text Book of Ore Dressing, 1940.
Richards, R.H. - Text Book of Ore Dressing, 1909.
Shaw, W.T. - Mining in the Lake Counties, 1970.
Wainwright, A. - A Pictorial Guide to the Lakeland Fells, 1964.

Future of Non-Ferrous Mining in Great Britain and Ireland, A Symposium, 1959.
Mines and Quarries General Report with Statistics, HMSO, 1914-1919 inc.
Mineral Survey Cumbria-Lakeland, 1932.
West Cumberland Industrial Survey, 1937.
Royal Commission on Metalliferous Mines and Quarries, HMSO, 1914.
The Premier Lamp and Engineering Company Ltd., Brochure.

BOOKS AVAILABLE FROM RED EARTH PUBLICATIONS

THE IRON MOOR

By ALEN McFADZEAN

For over 2,000 years the fells and valleys of Cumbria have been worked for the mineral riches entombed beneath them. Few sites have yielded more treasures than the mines of Lindal Moor where hematite — the blood red ore — was first exploited by the Cistercian monks of Furness Abbey.

Alen McFadzean unfolds the history and reveals the secrets of the Iron Moor from the advent of mining activity to its eventual demise in recent years. This is a story of masters and men, their struggles against the elements in a deep, dark, hostile environment, and their determination to survive in the face of natural and industrial crisis.

148 pages, illustrated with 16 plates plus maps and sketches.

WYTHBURN MINE AND THE LEAD MINERS OF HELVELLYN

By ALEN McFADZEAN

For forty years the miners of the Thirlmere valley tunnelled their way into the heart of Helvellyn — England's third highest mountain — in search of galena, the glittering, silver-rich lead ore of Cumberland's deep and elusive mineral lodes. Wythburn Mine was their place of work, a remote 19th century industrial settlement, steeped in Cornish mining tradition though embracing the techniques and expertise of the North Pennine ore fields.

Alen McFadzean recounts the fascinating history of Wythburn Mine, the men who laboured in it, and the managers who fought nature and financial hardship only to see their achievements torn down by Manchester Corporation and the execution of the Thirlmere Scheme, the controversial reservoir project of the 1870s and 1880s. This is a colourful, and sometimes tragic, account of the history of a vanished industry, the growth and decline of a mining community, and the flooding of a Lakeland Valley.

52 pages, illustrated with 12 plates plus maps and sketches.